T0321006

Innovative Strategies
in Tissue Engineering

RIVER PUBLISHERS SERIES IN RESEARCH AND BUSINESS CHRONICLES: BIOTECHNOLOGY AND MEDICINE

Volume 2

Series Editors

ALAIN VERTES
Sloan Fellow, London
Business School, Switzerland

PAOLO DI NARDO
Rome Tor Vergata, Italy

PRANELA RAMESHWAR
Rutgers University, USA

Combining a deep and focused exploration of areas of basic and applied science with their fundamental business issues, the series highlights societal benefits, technical and business hurdles, and economic potentials of emerging and new technologies. In combination, the volumes relevant to a particular focus topic cluster analyses of key aspects of each of the elements of the corresponding value chain.

Aiming primarily at providing detailed snapshots of critical issues in biotechnology and medicine that are reaching a tipping point in financial investment or industrial deployment, the scope of the series encompasses various specialty areas including pharmaceutical sciences and healthcare, industrial biotechnology, and biomaterials. Areas of primary interest comprise immunology, virology, microbiology, molecular biology, stem cells, hematopoiesis, oncology, regenerative medicine, biologics, polymer science, formulation and drug delivery, renewable chemicals, manufacturing, and biorefineries.

Each volume presents comprehensive review and opinion articles covering all fundamental aspect of the focus topic. The editors/authors of each volume are experts in their respective fields and publications are peer-reviewed.

For a list of other books in this series, www.riverpublishers.com
http://riverpublishers.com/series.php?msg=Research and Business Chronicles: Biotechnology and Medicine

Innovative Strategies in Tissue Engineering

Editors

Mayuri Prasad

Haematological Research Laboratory
Department of Haematology at
Aalborg University Hospital, Denmark

Paolo Di Nardo, MD

Laboratorio Cardiologia Molecolare & Cellulare
Dip. Scienze Cliniche e Medicina Traslazionale
Università di Roma Tor Vergata, Italy

River Publishers

Published, sold and distributed by:
River Publishers
Niels Jernes Vej 10
9220 Aalborg Ø
Denmark

ISBN: 978-87-93237-09-4 (Print)
        978-87-93237-10-0 (Ebook)

©2015 River Publishers

Contents

Preface

In spite of intensive investments and investigations carried out in the last decade, many aspects of the stem cell physiology, technology and regulation remain to be fully defined. After the enthusiasm that characterized the first decade of the discovery that, when given the right cue, stem cells could repair all the different tissues in the body; it is now time to start a serious and coordinated action to define how to govern the stem cell potential and to exploit it for clinical applications. This can be achieved only with shared research programs involving investigators from all over the world and making the results available to all.

The Disputationes Workshop series (http://disputationes.info) is an international initiative aimed at disseminating stem cell related cutting edge knowledge among scientists, healthcare workers, students and policy makers. The present book gathers together some of the ideas discussed during the third and fourth Disputationes Workshops held in Florence (Italy) and Aalborg (Denmark), respectively. The aim of this book is to preserve those ideas in order to contribute to the general discussion on organ repair and to bolster a fundamental scientific and technological leap towards the treatment of otherwise incurable diseases.

Editors

Pac

List of Abbreviations

3D	three dimensional
ABCB5	ATP-binding cassette sub-family B member 5
ABCG2	ATP-binding cassette sub-family G member 2
AIFA	Agenzia Italiana del farmaco (Italian Medicine Agency)
ALDH	Aldehyde dehydrogenase
AM	acrylamide
APA	aseptic processing area
APC	Antigen Presenting Cells
ATMP	Advanced Therapy Medicinal Products
BWP	Biologics Working Party
CAPS	cell aseptic processing system
CAT	Committee for Advanced Therapy
CeO_2	cerium dioxide
CHMP	Committee for Human Medicinal Products
CKI	Casein kinase 1
CPCs	cardiac progenitor cells
CPF	cell processing facility
CPWP	Cell Products Working Party
CSC	Cancer stem cells
CTMP	Cell Therapy Medicinal Products
DKK-1	Dickkopf-related protein -1
DMAA	N,N-dimethylacrylamide
DPTE	Double Porte de Transfert Etanche
EATRIS- ERIC	European Advanced Translational Research Infrastructure-European Research Infrastructure Consortium
EB	embryoid body
ECM	extracellular matrix
EMA/EMEA	European Medicine Agency
ESC	embryonic stem cell
EU	European Union
EU-CJ	European Court of Justice
FACS	Fluorescence-activated cell sorting

Materials developed in this sense are important for drug delivery systems and tissue engineering approaches and can play a key role in developing of artificial organs or on growing organs.

Many types of polymers can be used as carrier systems due to their ability to provide delivery of active substances to specific sites. Moreover, such biomaterials can deliver cells to the surrounding tissue, which makes them excellent candidates for controlling the attachment, growth and differentiation of the cells [1].

Bioactive ceramics can form a mechanically strong interfacial bond with bone depending on the conditions. Thus, bioactive composites present excellent biochemical compatibility, but less optimal biomechanical compatibility. To be an ideal bone replacement material, composite biomaterials have to combine bioactivity with biomechanical properties [2].

Biomaterials can find applications in biomedicine as soft or hard implants and can be used as: joint replacements, bone cements, artificial ligaments, dental implants, blood vessels, heart valves, skin repair, contact lenses, and cochlear replacement [3].

Considering the various types of implants that can be engineered and their functions, they need to be designed with appropriate geometry, size and weight for a given patient. To obtain tailored made prosthesis, many researchers have incorporated different features to promote tissue ingrowth [4], several described rapid prototyping [5, 6] to create artificial tissue by means of computer numerical controlled machining and others used electron beam melting [7] to fabricate complex shape implants. In the future years it is overseen a dramatically increase in use of biomaterials and development of advanced materials in medical-device industry due to materials complex properties and shapes [8].

At the beginning, the only requirement for materials when they were used for the first time in biomedical applications was to be "inert" so as to reduce the inflammatory response [9]. These type of materials were classified as "first – generation" and had to possess proper combination of physical properties as for the replaced tissue with a low toxic response of the body.

The second-generation of biomaterials appeared between 1980 and 2000. They were completely opposite to the first generation in terms of interaction to the human body. These materials had the ability to interact with the biological environment, to improve their biointegration and some were bioabsorbable; they undergo degradation while the new tissue was regenerated.

The third-generation of biomaterials combines bioactivity and biodegradability with the ability to stimulate cellular response [10]. Moreover, three-dimensional porous structures are being developed that can stimulate cell proliferation or can act as drug delivery systems [11–13]. Tissue engineering and the third-generation of biomaterials appeared approximately at the same time as a potential solution to tissue transplantation and grafting [14–16]. Regenerative medicine is a recent research area, which explores ways to repair and regenerate organs and tissues using combination of stem cells, growth factors and peptide sequences with synthetic scaffolds [17, 18].

1.2 Factors that Influence the Quality of a Biomaterial

Polymer matrices used in therapeutic applications are often resorbed or degraded in the body. In this case, two key challenges can be identified. First, the degradable polymers used in tissue engineering were selected from materials used for other surgical uses and thus such materials may have deficiencies in terms of mechanical and degradation properties [19]. To overcome this, new classes of polymers and biopolymers are being developed. The second major challenge concerns tailoring these polymers into scaffolds with defined and complex porous shapes, which can undergo cell attachment and proliferation [20, 21].

Biodegradable three-dimensional scaffolds play an important role in maintaining the cell functions. The cells adhere to the porous scaffold in all three directions, proliferate and replace the temporary scaffold. Moreover, the scaffold should be biodegradable, biocompatible, and highly porous with a large surface area, with a specific mechanical strength and shape to permit cell attachment, proliferation and maintaining of differentiated cell functions [22].

When obtaining a biomaterial several factors need to be considered starting with selection of raw materials in terms of purity, toxicity and biodegradability, the obtaining process and last but not least the mechanical and structural properties of the final material.

Polymer purity as well as purity of other additives is important for biocompatibility of medical devices and drug delivery systems. Residual monomers, catalyst residue or impurities strongly affect cell viability; a special consideration must be given also to the biodegradability compounds. Therefore, quality assurance of productions processes must be taken into account especially for biocompatibility of newly obtained biomaterials [23].

Many natural polymers have been extensively used in biomaterials in tissue repair and regeneration, for example collagen, gelatine and

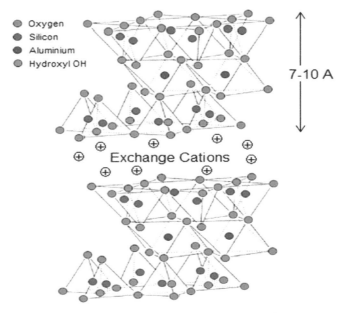

Figure 1.1   Montmorillonite structure (Modified form [34])

Montmorillonite can absorb water increasing the volume of 20–30 times. Each layered sheet is smaller than 1 nm thick, with surface of about 1 µm (1000 nm). For an aspect ratio (length to thickness ratio of silicate layers) of about 1000, clays specific surface area is about 750 m^2/g, resulting in high reinforcement efficiency at lower clay concentrations (2–5%) [39]. However, natural layered silicates are not suitable for obtaining nanocomposites because they are too hydrophilic and layers are compacted too tightly by inorganic cations to interact with the hydrophobic molecules of polymer and to be able to disperse among them. To ensure compatibility between the clay and the polymer matrix, modification of the layered silicate surface (organophilisation) is necessary. This technology implies two steps:

- layered silicate purification to a sufficiently high level requested by the field of application;
- modification of the layered silicate surface (organophilisation).

The large specific surface area, layer charge, swelling capacity and adsorption properties of different organic/ inorganic substances make clay minerals (hydrated layered silicates) to be useful as materials for pollution control [40],

geological deposits consists in replacing the exchangeable cations with Na^+ followed by washing with water.

The natural layered silicates (bentonites) may contain different percentages (5–40%) of quartz and other impurities, which act as a sterile, hindering the surface modification process, in order to ensure the compatibility between the silicate and a polymer matrix [48]. Purification is made by dispersing the layered silicate in distilled water at 60–90 °C. Quartz and other hydrophobic impurities are separated by decantation [49]. Often is required additional purification to ensure a high degree of purity, imposed in the biomedical field. By additional purification, the concentration of MMT increases with 7–10% [49].

1.3.3 Layered Silicate in Drug Release Systems

Layered silicates are widely used ingredients in pharmaceutical products as both excipients and active substances. Based on their adsorption capacities, swelling and colloidal properties, clay minerals can be used in drug delivery systems to achieve technological (taste masking), chemical (increasing stability), biopharmaceutical (decreasing or increasing dissolution rate, delaying and/or targeting drug release) and pharmacological (prevention or reduction of side effects) benefits [50].

Layered silicates, especially montmorillonite and saponite, because of their high cation exchange capacity, were used and studied in controlled-release drug delivery systems. The interaction between clay minerals and active substances depends on the type of mineral involved and on the functional groups and the properties of the organic compounds.

There are some mechanisms of interaction or complexation between montmorillonite and drug [51–67]:

- Cation exchange with cationic drugs. This produces a strong interaction bonds between montmorillonite and basic molecules;
- Anion exchange of anionic drugs at slightly positive-charged platelet edges. This produces weak interaction bonds with anionic drugs;
- Hydrogen bonding at platelet faces;
- Intercalation of non-ionic drugs via ion-dipole interactions;
- Adsorption by solvent deposition onto the high surface area of the clay to increase the dissolution rate of poorly soluble drugs.

Depending on the degree of interaction between montmorillonite and drug, nanostructured systems with intercalated, partial exfoliated or exfoliated lamellar structures, able to release faster or slowly the bioactive substance could be obtained [68].

1.3.4 Biopolymers Properties

Biopolymers are materials produced from renewable resources. In recent years, the worldwide interest in biopolymers increased due to their positive environmental impact such as reduced carbon dioxide emissions. Many biopolymers are biodegradable, being degraded and gradually absorbed and/or eliminated by the body. This property is of high interest for biomedical applications (tissue engineering, bone replacement/repair, dental applications and controlled drug delivery).

Many biomedical applications require biomaterials with high performance and mechanical properties. For such materials to be used in biomedicine is not enough to be biocompatible and biodegradable, certain mechanical properties are imposed (i.e., low friction coefficient, wear resistance, thermal stability, modulus, strength and toughness). Not all these properties can be achieved by using the biopolymer alone [69–71]. By dispersion of inorganic/organic fillers at the nanometer scale into a biopolymer matrix, new class of bionanocomposites, with enhanced mechanical properties as compared to conventional microcomposites, was developed.

One of the most researched and used biomaterial in various fields of medicine is collagen, due to its biocompatibility, biodegradability, and weak antigenicity [72]. It is well known the use of collagen as biomaterial, biocompatible and bioresorbable for connective tissue prosthesis in which collagen is the basic protein. To use collagen as a scaffold in bone reconstruction, modifications are necessary in the structure and composition of the matrix to achieve the osteoconductive and osteoinductive effect. This was achieved by preparation of biocomposites with SiO_2, TiO_2, clays, hydroxyapatite, etc. [73, 74]. The collagen fibrils have high elasticity while the mechanical properties are relatively limited. Substantial improvement of its properties can be achieved by the nanoscale dispersion of the layered silicate in a collagen matrix. Depending on the collagen morphologies, nanocomposites with intercalated or exfoliated lamellar structures and improved thermal stability were obtained [75].

In order to use a material in biomedicine, both its bulk and surface properties are important to be known, especially interfacial behaviour with aqueous environment. The wettability capacity, the swelling behaviour, the presence of surface roughness, liquid and vapour water absorption are only

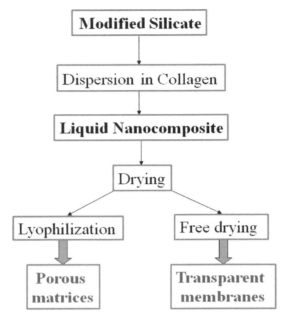

Figure 1.3   Scheme flow for obtaining collagen based nanocomposites

dimensional porous scaffolds can be obtained from synthetic polymers such as poly(glycolic acid) (PGA), poly(lactic acid) (PLA), and their copolymer poly(DL- lactic-co-glycolic acid) (PLGA), and from naturally derived polymers such as collagen [86–88].

To obtain hybrid collagen microsponges, gaseous glutaraldehyde was used as cross-linking agent to form their pores and to stabilize the collagen in water [22, 89, 90].

1.5 Biomaterials Development

The quality of biomaterials is strongly influenced by composition, architecture and three-dimensional design, biocompatibility, but also by mechanical strength of the scaffold that mimics the mechanical strength of the tissue intended to repair or replace. Pore size distribution as well as pore types influences the attachment of specific cells and interaction of biomaterials with the body. Furthermore, it is important also to identify and isolate the appropriate cells from the primary source, when selecting the cells for the engineered scaffold [91].

Methylene blue membrane

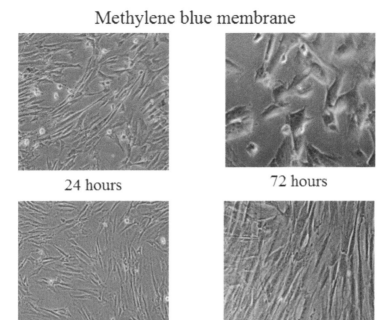

24 hours   72 hours

Quebracho membrane

Figure 1.5   Microscopic analysis of human dermal fibroblast culture samples sowed on collagen/clay/bioactive substance membranes (Reproduced with permission from [92])

on human dermal fibroblast (HDF) cells. Figure 1.5. presents *in vitro* results after 24 and 72 hours, respectively. The cells were uniformly distributed and presented a normal phenotype after 24 h. However, quebracho membrane showed a 99% cell viability as compared to methylene blue membrane which after 72 h presented only 30% viable cells [92].

Water vapour adsorption was also tested for the same type of materials and compared to collagen membrane. Figure 1.6. shows the adsorption curves of the membranes for 48 hours. It was observed that the collagen membrane presented a continuous increasing variation as compared to nanocomposite membranes. The hybrids continuously adsorbed water vapours in the first 24 hours from exposure, then reached a plateau and the adsorption remained constant [92].

These findings showed that such type of materials especially the membrane with quebracho, which presented a biostimulating effect on the growth and development of fibroblast cells, could be used as antiseptic and good regenerating patches [92].

Other collagen/layered silicate nanocomposites which contained gentamicine as active substance were obtained. *In vitro* biocompatibility test was performed on human dermal fibroblast cells and the results were compared to a collagen membrane as it can be seen in Figure 1.7. The layered silicate increased the biocompatibility of the materials; the cells presented a normal phenotype and proliferated. As compared to the collagen

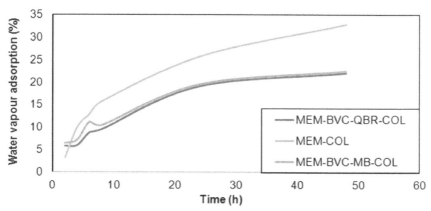

Figure 1.6 Water vapour adsorption for nanocomposite membranes with quebracho and methylene blue (Reproduced with permission from [92])

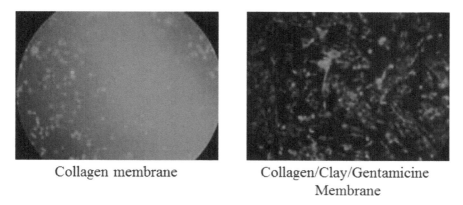

Collagen membrane   Collagen/Clay/Gentamicine Membrane

Figure 1.7 Microscopic analysis of human dermal fibroblast culture samples sowed on collagen/clay/gentamicine membrane (Reproduced with permission from [74])

[5] Y. He, M. Ye, C. Wang, 'A method in the design and fabrication of exact-fit customized implant based on sectional medical images and rapid prototyping technology', Int. J. Adv. Manuf. Technol., pp. 504–508, 28, 2006.

[6] T. Burg, C. A. P. Cass, R. Groff, M. Pepper, K. J. L. Burg, 'Building off-the-shelf tissue engineered composites', Phil. Trans. R. Soc. A, pp. 1839–1862, 368, 2010.

[7] L. E. Murr, S. M. Gaytan, F. Medina, H. Lopez, E. Martinez, B. I. Machado, D. H. Hernandez, L. Martinez, M. I. Lopez, R. B. Wicker, J. Bracke, 'Next-generation biomedical implants using additive manufacturing of complex, cellular and functional mesh arrays', Phil. Trans. R. Soc. A, pp. 1999–2032, 368, 2010.

[8] R. J. Narayan, 'The next generation of biomaterial development', Phil. Trans. R. Soc. A, pp. 1831–1837, 368, 2010.

[9] L. L. Hench, 'Biomaterials', Science, pp. 826–831, 208, 1980.

[10] L. L. Hench, J. Polak, 'Third generation biomedical materials', Science, pp. 1014–1017, 295, 2002.

[11] D. Hutmacher, M. B. Hurzeler, H. Schliephake, 'A review of material properties of biodegradable and bioresorbable polymer for GTR and GBR', J. Oral Maxillofac. Implants, pp. 667–678, 11, 2000.

[12] J. S. Temenoff, A. G. Mikos, 'Tissue engineering for regeneration of articular cartilage', Biomaterials, pp. 431–440, 21, 2000.

[13] C. M. Agrawal, R. B. Ray, 'Biodegradable polymeric scaffolds for musculoskeletal tissue engineering', J. Biomed. Mater. Res., pp. 141–150, 55, 2001.

[14] J. C. Fernyhough, J. J. Schimandle, M. C. Weigel, C. C. Edwards, A. M. Levine, 'Chronic donor site pain complicating bone graft harvest from the posterior iliac crest for spinal fusion', Spine, pp. 1474–1480, 17, 1992.

[15] J. C. Banwart, M. A. Asher, R. S. Hassanein, 'Iliac crest bone graft harvest donor site morbidity. A statistical evaluation', Spine, pp. 1055–1060, 20, 1995.

[16] J. A. Goulet, L. E. Senunas, G. L. DeSilva, M. L. V. H. Greengield, 'Autogenous iliac crest bone graft. Complications and functional assessment', Clin. Orthop., pp. 76–81, 339, 1997.

[17] P. Hardouin, K. Anselme, B. Flautre, F. Bianchi, G. Bascoulenguet, Bouxin, B. 'Tissue engineering and skeletal diseases', Joint Bone Spine, pp. 419–424, 67, 2000.

[31] K. A. Athanasiou, G. G. Niederauer, C. M. Agrawal, 'Steriliza-tion, toxicity, biocompatibility and clinical applications of polylactic acid/ polyglycolic acid copolymers', Biomaterials, pp. 93–102, 17, 1996.

[32] L. G. Griffith, 'Polymeric biomaterials', Acta mater., pp. 263–277, 48, 2000.

[33] S. H. Teoh, 'Fatigue of biomaterials: a review', International Journal of Fatigue, pp. 825–837, 22, 2000.

[34] K. Magniez, 'Development of novel melt-spun nanocomposite fibers', Society of Plastics Engineers, pp. 1–3, 10.1002/spepro.003802, 2011.

[35] O. Solacolu, 'Physical chemistry of technical silicates', Second Edition, Technical Publishing, Bucharest, 1968.

[36] J. C. Hutchison, R. Bissessur, D. F. Shriver, 'Enhancement of ion mobility in alumino-silicate-polyphosphazene nanocomposites', Mat. Res. Soc. Symp. Proc., pp. 489–494, 457, 1997.

[37] R. A. Vaia, E. P. Giannelis, 'Lattice model of polymer melt inter-calation in organically-modified layered silicates', Macromolecules, pp. 7990–7999, 30, 1997.

[38] B. K. G. Theng, 'Formation and properties of clay - polymer complexes', Elsevier Scientific Publishing Co., Amsterdam, Oxford, New York, 1979.

[39] R. L. D'Aquino, 'A little clay goes a long way', Chem. Eng. Mag., pp. 1–2, 106, 1999.

[40] G. J. Churchman, W. P. Gates, B. K. G. Theng G. Yuan, 'Clays and clay minerals for pollution control', Handbook of Clay Science, pp. 625–675, Elsevier, Oxford, UK, 2006.

[41] S. Nir, Y. El Nahhal, T. Undabeytia, G. Rytwo, T. Polubesova, Y. Mishael, U. Rabinovitz, B. Rubin, 'Clays and pesticides', Handbook of Clay Science, pp. 677–691, Elsevier, Oxford, UK, 2006.

[42] K. Czurda, 'Clay liners and waste disposal', Handbook of Clay Science, pp. 693–701, Elsevier, Oxford, UK, 2006.

[43] R. Pusch, 'Clays and nuclear waste management', Handbook of Clay Science, pp. 703–716, Elsevier, Oxford, UK, 2006.

[44] M. I. Carretero, C. S. F. Gomes, F. Tateo, 'Clays and human health', Handbook of Clay Science, pp. 717–741, Elsevier, Oxford, UK, 2006.

[45] M. T. Droy-Lefaix, F. Tateo, 'Clays and clay minerals as drugs', Handbook of Clay Science, pp. 743–752, Elsevier, Oxford, UK, 2006.

[46] M. Ross, R. P. Nolan, A. M. Langer, W. C. Cooper, 'Health effects of mineral dusts other than asbestos', Health Effects of Mineral Dusts.

Reviews in Mineralogy, pp. 361–409, 28, Mineralogical Society of America, Washington, DC, 1993.

[47] Y. C. Ke, P. Stroeve 'Polymer-layered silicate and silica nanocomposites', Elsevier B. V., 2005.

[48] K. A. Carrado, 'Synthetic clay minerals and purification of natural clays', Handbook of Clay Science, pp. 115–139, Elsevier, Oxford, UK, 2006.

[49] Z. Vuluga, G. C. Potarniche, M. G. Albu, V. Trandafir, D. Iordachescu, E. Vasile, 'Collagen - modified layered silicate biomaterials for regenerative medicine of bone tissue', Materials Science and Technology, pp. 125–148, InTech, 2012.

[50] C. Aguzzi, P. Cerezo, C. Viseras, C. Caramella, 'Use of clays as drug delivery systems: Possibilities and limitations', Applied Clay Science, pp. 22–36, 36 (1–3), 2007.

[51] Y. Chen, A. Zhou, B. Liu, J. Liang, 'Tramadol hydrochloride/ montmorillonite composite: Preparation and controlled drug release', Appl. Clay Sci., pp. 108–112, 49 (3), 2010.

[52] M. S. Lakshmi, M. Sriranjani, H. B. Bakrudeen, A. S. Kannan, A. B. Mandal, B. S. R. Reddy, 'Carvedilol/montmorillonite: Processing, characterization and release studies', Appl. Clay Sci., pp. 589–593, 48 (4), 2010.

[53] G. V. Joshi, B. D. Kevadiyaa, H. A. Patel, H. C. Bajaj, R. V. Jasrab, 'Montmorillonite as a drug delivery system: Intercalation and *in vitro* release of timolol maleate', Int. J. Pharm., pp. 53–57, 374, 2009.

[54] N. Meng, N. Zhou, S. Zhang, J. Shen, 'Controlled release and antibacterial activity chlorhexidine acetate (CA) intercalated in montmorillonite', Int. J. Pharm., pp. 45–49, 382 (1–2), 2009.

[55] J. K. Park, Y. B. Choy, J.-M. Oh, J. Y. Kima, S.-J. Hwanga, J. H. Choy, 'Controlled release of donepezil intercalated in smectite clays', Int. J. Pharm., pp. 198–204, 359, 2008.

[56] J. P. Zheng, L. Luan, H. Y. Wang, L. F. Xi, K. D. Yao, 'Study on ibuprofen/montmorillonite intercalation composites as drug release system', Appl. Clay Sci., pp. 297–301, 36, 2007.

[57] T. Kollár, I. Palinko, Z. Konya, I. Kiricsi, 'Intercalating amino acid guests into montmorillonite host', J. Mol. Struct., pp. 335–340, 651–653, 2003.

[58] F. H. Lin, Y. H. Lee, C. H. Jian, 'A study of purified montmorillonite intercalated with 5-fluorouracile as drug carrier', Biomaterials, pp. 1981–1987 23 (9), 2002.

culture in biodegradable polymer scaffolds', J. Biomed. Mater. Res., pp. 17–28, 36, 1997.

[84] M. G. Dunn, J. B. Liesch, M. L. Tiku, J. P. Zawadsky, 'Development of fibroblast-seeded ligament analogs for ACL reconstruction', J. Biomed. Mater. Res., pp. 1363–1371, 29, 1995.

[85] G. Nechifor, S. I. Voicu, A. C. Nechifor, S. Garea, 'Nanostructure hybrid membrane polysulfone-carbon nanotubes for hemodyalisis', Desalination, pp. 342–348, 241, 2009.

[86] S. J. Peter, M. J. Miller, A. W. Yasko, M. J. Yaszemski, A. G. Mikos, 'Polymer concepts in tissue engineering', J. Biomed. Mater. Res., pp. 422–427, 43, 1998.

[87] L. E. Freed, G. Vunjak-Novakovic, R. J. Biron, D. B. Eagles, D. C. Lesnoy, S. K. Barlow, R. Langer, 'Biodegradable polymer scaffolds for tissue engineering', Bio/Technology, pp. 689–693, 12, 1994.

[88] B. S. Kim, D. J. Mooney, 'Development of biocompatible synthetic extracellular matrices for tissue engineering', TIBTECH, pp. 224–230, 16, 1998.

[89] L. H. H. Olde Damink, P. J. Dijkstra, M. J. A. van Luyn, P. B. van Wachem, P. Nieuwenhuis, J. Feijen, 'Glutaraldehyde as a crosslinking agent for collagen-based biomaterials', J. Mater. Sci., pp. 460–472, 6, 1995.

[90] N. Barbani, P. Giusti, L. Lazzeri, G. Polacco, G. Pizzirani, 'Bioartificial materials based on collagen: 1. Collagen cross-linking with gaseous glutaraldehyde', J. Biomater. Sci., pp. 461–469, 7, 1995.

[91] M. S. Chapekar, 'Tissue Engineering: Challenges and Opportunities', J Biomed Mater Res (Appl Biomater), pp. 617–620, 53, 2000.

[92] C.-G Potarniche, Z. Vuluga, C. Radovici, S. Serban, D. M. Vuluga, M. Ghiurea, V. Purcar, V. Trandafir, D. Iordachescu, M. G. Albu, 'Nanocomposites based on collagen and Na-montmorillonite modified with bioactive substances', Materiale Plastice, pp. 267–273, 47, 2010.

pulse, CPCs were able to take up the NPs and retain them inside the cytosol, while preserving their stemness phenotype and multipotential capability at all time-points considered. Moreover, when challenged with 50 µM H_2O_2 for 30 min, CeO_2-treated CPCs were protected from the oxidative stress. In particular, after 24 h, only the highest concentration was protective; after 7 d, ROS levels were mitigated with all concentrations. This study demonstrated that internalized CeO_2 NPs can act as a long-term defense against the oxidative insult. NPs were activated only when cells were hit by an external oxidative perturbation, remaining inert in respect to the main CPC characteristics. In conclusion, these results suggest that CeO_2 nanoparticles hold an enormous potential in TE treatments protecting stem cells against the oxidative damage.

Keywords: Cerium dioxide, cardiac precursor cells, tissue engineering, reactive oxygen species.

2.1 Interaction of Cerium Oxide Nanoparticles with Biological Systems

Over the last few years, nanotechnology has made significant strides especially in the field of regenerative medicine, thus enabling the development of a new generation of nanostructured biomaterials for medical applications. In particular, nano-composite hybrid scaffolds, made by incorporating nanoparticles into bio-compatible/erodible polymeric matrices, have gained rising attention. The possibility to fine-tune the properties of these materials to meet a broad range of applications makes them attractive systems for tissue engineering. For example, polymeric scaffolds loaded with hydroxyapatite nanoparticles are already used for bone tissue reconstruction [1]. In this respect, deciphering how cells interact with scaffolds and the mechanisms through which nano-components are internalized without exerting direct effects on cell behavior is particularly intriguing in order to obtain novel biomaterials with promising and controllable properties.

Recently, cerium oxide nanoparticles (CeO_2, nanoceria) have been demonstrated to favor cardiac precursor cell (CPC) adhesion and growth when embedded into PLGA scaffolds [2]. In particular, cerium oxide nanoparticle filling of PLGA films resulted in enhanced mechanical properties and in a change in scaffold nano-rugosity. On these functionalized supports, cells exhibited better adhesion and growth as compared with PLGA alone. CPCs were able to acquire a typical alignment, due to support rugosity, which, combined with that determined by the presence of ceramic nanoparticles, provided better anchorage sites for cell engraftment. Nevertheless, cardiac-derived cells displayed better growth performance when cultured onto CeO_2-PLGA films, as compared with films loaded with titanium oxide (TiO_2), thus suggesting a potential chemical stimulus can be exerted by ceria nanoparticles on cardiac resident progenitor cells [2].

CeO_2 is a rare earth oxide material of the lanthanide series commonly used in important industrial applications [3, 4], but recent reports highlighted the beneficial effects of cerium oxide in biological systems [5, 6]. In particular, it has been proposed that ceria nanoparticles could exhibit an oxidant scavenging activity reducing the cytotoxic effects of intracellular oxidative stress conditions via changes of the oxidation state: Ce^{4+} / Ce^{3+} [7–9]. Ceria nanoparticles display their unique property to store and release oxygen because of the great mobility of these atoms inside the lattice; each released oxygen atom causes the formation of a vacancy and electron transfer to Ce^{4+} which reduces to Ce^{3+}. This mechanism seems to be greatly facilitated in nanoparticles, where the higher surface area is accompanied by more oxygen vacancies and, thus, higher Ce^{3+} concentration in the lattice, resulting in enhanced catalytic properties [10, 11]. Indeed, reactive oxygen species (ROS), such as superoxides and peroxides, could react on these active sites and be counteracted; as a consequence, Ce^{3+} ions would be oxidized in Ce^{4+} ions in a reversible and autocatalytic way. This is because of the cerium ability to switch between the 3+ state under reducing conditions and 4+ state under oxidizing conditions [6, 12]. This ability, combined with multiple active sites that may be generated on a single nanoparticle, could provide an explanation to ceria antioxidant properties with the ability to scavenge ROS and mostly as a catalyst with superoxide dismutase (SOD) and catalase mimetic activities [13–15]. These properties candidate ceria as a novel long-lasting antioxidant compound with the promise to actively participate in mitigating oxidative stress, which is considered a critical actor in the establishment and progression of several diseases, including cardiovascular dysfunctions [16–19], or after treatments such as chemotherapy [20].

2.2 Cerium Oxide Nanoparticles Shield Cardiac Precursor Cells against the Oxidative Stress

In the last decade, evidence has been acquired that an adult stem cell pool is present in almost every organ of the body. These cells are endowed with self-renewal capability and can be committed to a specific cell lineage. The identification of a cardiac progenitor cell (CPC) population in the adult

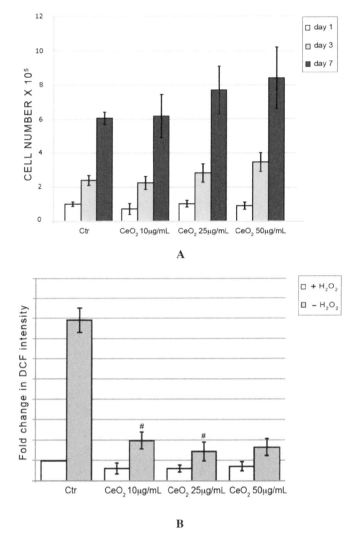

Figure 2.3 A) CPC proliferation assessed at 1 d, 3 d, and 7 d after 24 h CeO_2 exposure. The values are expressed as means $\pm$ SD of three independent experiments. (#= p > 0.05). B) Effect of H2O2 on intracellular ROS levels in Lin- Sca-1pos CPCs at 7 days after CeO_2 treatment. ROS production, measured using a DCFH probe, decreased with all NPs concentrations tested. (#= H_2O_2 treated cells vs. CeO_2-H2O2 treated cells). The values are expressed as means $\pm$ SD of the fold change in DCF fluorescence intensity with respect to H_2O_2-untreated control (ctr-) from three different tests (p < 0.05).

effects were not detectable in other cell types [41]. Therefore, it could be worth conducting further investigations in order: i) to elucidate the biological mechanisms behind the action of cerium oxide; ii) to understand the interactions between this promising material and tissues (both healthy and damaged) *in vivo*.

References

[1] N. T. Ba Linh, Y. K. Min, B. T. Lee, 'Hybrid hydroxyapatite nanoparticles-loaded PCL/GE blend fibers for bone tissue engineering' J Biomater Sci Polym Ed., 24:520–38 doi: 10.1080/09205063.2012. 697696, 2013.

[2] C. Mandoli, F. Pagliari, S. Pagliari, G. Forte, P. Di Nardo, S. Licoccia, E. Traversa, 'Stem Cell Aligned Growth Induced by CeO_2 Nanoparticles in PLGA Scaffolds with Improved Bioactivity for Regenerative Medicine' Adv. Funct. Mater., 20, 1617–1624, 2010.

[3] A. M. El-Toni, S. Yin, T. Sato, 'Enhancement of Calcia Doped Ceria Nanoparticles Performance as UV Shielding Material' Adv. Sci. Technol., 45, 673–678, 2006.

[4] V. Esposito, E. Traversa, 'Design of electroceramics for solid oxide fuel cell applications: playing with ceria' J. Am. Ceram. Soc., 91, 1037–1051, 2008.

[5] J. Chen, S. Patil, S. Seal, J. F. McGinnis, 'Rare Earth Nanoparticles Prevent Retinal Degeneration Induced by Intracellular Peroxides' Nat. Nanotechnol., 1, 142–150, 2006.

[6] A. S. Karakoti, N. A. Monteiro-Riviere, R. Aggarwal, J. P. Davis, R. J. Narayan, W. T. Self, J. McGinnis, S. Seal, 'Nanoceria as Antioxidant: Synthesis and Biomedical Applications' JOM (1989) 60, 33–37, 2008.

[7] R. W. Tarnuzzer, J. Colon, S. Patil, S. Seal, 'Vacancy Engineered Ceria Nanostructures for Protection from Radiation-Induced Cellular Damage' Nano Lett., 5, 2573–2577, 2005.

[8] F. Esch, S. Fabris, L. Zhou, T. Montini, C. Africh, P. Fornasiero, G. Comelli, R. Rosei, 'Electron Localization Determines Defect Formation on Ceria Substrates' Science, 309, 752–755, 2005.

[9] I. Celardo, J. Z. Pedersen, E. Traversa, L. Ghibelli, 'Pharmacological potential of cerium oxide nanoparticles' Nanoscale, 3:1411–20. doi: 10.1039/c0nr00875c, 2011.

[10] A. Migani, G. N. Vayssilov, S. T. Bromley, F. Illas, K. M. Neyman, 'Greatly Facilitated Oxygen Vacancy Formation in Ceria Nanocrystallites' Chem. Comm., 46, 5936–5938, 2010.

[11] T. C. Campbell, C. H. Peden, 'Chemistry. Oxygen Vacancies and Catalysis on Ceria Surfaces' Science, 309, 713–714, 2005.

[12] M. Das, S. Patil, N. Bhargava, J. F. Kang, L. M. Riedel, S. Seal, J. J. Hickman, 'Auto-Catalytic Ceria Nanoparticles Offer Neuroprotection to Adult Rat Spinal Cord Neurons' Biomater., 28, 918–1925, 2007.

[13] E. G. Heckert, A. S. Karakoti, S. Seal, W. T. Self, 'The Role of Cerium Redox State in the SOD Mimetic Activity of Nanoceria' Biomater., 29, 2705–2709, 2008.

[14] C. Korsvik, S. Patil, S. Seal, W. T. Self, 'Superoxide Dismutase Mimetic Properties Exhibited by Vacancy Engineered Ceria Nanoparticles' Chem. Comm., 14, 1056–1058, 2007.

[15] T. Pirmohamed, J. M. Dowding, S. Singh, B. Wasserman, E. Heckert, A. S. Karakoti, J. E. King, S. Seal, W. T. Self, 'Nanoceria Exhibit Redox State-Dependent Catalase Mimetic Activity' Chem. Comm., 46, 2736–2738, 2010.

[16] R. Kohen, A. Nyska, 'Oxidation of Biological Systems: Oxidative Stress Phenomena, Antioxidants, Redox Reactions, and Methods for Their Quantification' Toxicol. Pathol., 30, 620–650, 2002.

[17] B. Kumar, S. Koul, L. Khandrika, R. B. Meacham, H. K. Koul, 'Oxidative Stress Is Inherent in Prostate Cancer Cells and Is Required for Aggressive Phenotype' Cancer Res., 68, 1777–1785, 2008.

[18] M. K. Misra, M. Sarwat, P. Bhakuni, R. Tuteja, N. Tuteja, 'Oxidative Stress and Ischemic Myocardial Syndromes' Med. Sci. Monit., 15, 209–219, 2009.

[19] G. S. Gaki, A. G. Papavassiliou, 'Oxidative Stress-Induced Signaling Pathways Implicated in the Pathogenesis of Parkinson's Disease' Neuromolecular Med., 2014.

[20] K. A. Conklin, 'Chemotherapy-associated oxidative stress: impact on chemotherapeutic effectiveness' Integr Cancer Ther., 3:294–300, 2004.

[21] K. Urbanek, D. Torella, F. Sheikh, A. De Angelis, D. Nurzynska, F. Silvestri, C. A. Beltrami, R. Bussani, A. P. Beltrami, F. Quaini, R. Bolli, A. Leri, J. Kajstura, P. Anversa, 'Myocardial Regeneration by Activation of Multipotent Cardiac Stem Cells in Ischemic Heart Failure' Proc. Natl. Acad. Sci. USA., 102, 8692–8697, 2005.

[22] O. Bergmann, R. D. Bhardwaj, S. Bernard, S. Zdunek, F. Barnabé-Heider, S. Walsh, J. Zupicich, K. Alkass, B. A. Buchholz, H. Druid, S. Jovinge,

model of the lowest possible phylogenic scale and the lowest possible animal numbers is encouraged [1]. On the other hand the researcher must consider the translational relevance of the animal species and use an appropriate number of animals to ensure significance. It is also important to conduct pilot experiments where variation can be tested and this data can be used for power calculations. Experiments with too low animal numbers do not save animals as the ones used go to waste and in the end a higher number of animlas will be sacrificed to obtain reliable results. Pilot studies are also very valuable for optimizing methods, e.g. surgical skills. It is important to achieve a certain level of reproducibility in an animal model because a high variation or mortality will require higher animal numbers. All these efforts are qualified as refinements. Finally, before starting an animal experiment, it is important to fully explore all *in vitro* test methods to fulfil the replacement demand. A large number of biocompatibility and *in vitro* characterisation studies can be done using cell culture or organotypic cultures.

In this chapter, we will briefly review animal testing in general and show how we complement animal testing with *in vitro* organ equivalent models for testing of biomaterials, using the development of biosynthetic corneal implants as an example.

3.2 Designing Animal Experiments

3.2.1 Randomization and Blinding

The lack of a proper randomization protocol is an important source of bias. Many animal studies do not report the randomization method or even whether it was used. Picking animals randomly from a cage does not qualify as randomization. The physiological status of the animal can affect the ease at which it is caught thus producing bias [2]. One must develop a system where the animal selection is totally at chance, e.g. using sealed envelopes or throwing dice. Both complete randomization and randomized blocking can be used [3]. Another important factor is blinding the experimental groups, so that the person analysing the data and evaluating outcomes does not know which animals received the treatment and which the controls are. This ensures that the researchers do not tend to see results that they expect or hope for. This is especially important when using subjective criteria or scoring systems for outcomes. Generally, these should be avoided if possible and substituted by more objective ones, e.g. instead of mild to severe redness, a number for the area affected or intensity of colour change [4].

3.2.2 Control Groups

Designing the appropriate control groups is crucial for obtaining reliable results. They should include positive, negative, sham and comparative controls at the least. When performing any invasive, traumatic procedure it is important to have sham-operated animals to test whether the procedure itself could affect the outcome. In the case of developing new treatments as an alternative to established ones, it is desirable to include a group receiving the standard treatment as a comparative control [5].

3.2.3 Statistical Analysis

The selection of the appropriate statistical test is extremely important for the outcome of the study. It should be selected already at the planning stage based on the type of data obtained, expected distribution and variation. It is not acceptable to analyse the experimental data with a series of tests until one of them generates the desired p-value. The statistical analysis method should be decided upon after a pilot study or based on data obtained from published studies along with the power calculations and estimation of the minimum number of animals needed.

3.2.4 Design Stages

Any animal experiment should be carefully planned and preceded by a thorough literature search to obtain as much information as possible about the available animal models and their limitations as well as to avoid repeating experiments that have already been done. Performing pilot studies is very beneficial, especially for new experimental systems, as they can lead to refinement of the methodology and help calculate the group sizes. As mentioned above, selection of the statistical method should be done at this stage [6].

3.3 Limitations of Animal Models

There have been many concerns raised about the usefulness of animal experimentation to predict the outcome of human responses to treatments or assess their safety.

3.3.1 Animal Species

As mentioned above the demand to choose the animal species on the lowest phylogenic level has to be balanced with selecting the one closest to humans with regard to the process or organ to be studied. There have been some

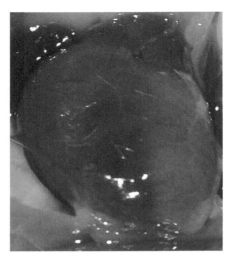

Figure 3.1 Ischemia resulting from ligation of the LAD in a rat model.

An overview of human and animal trials using stem cell therapies for heart regeneration showed that large animal myocardial infarct models do exhibit clinical relevance [10]. The results from rodent models are unfortunately not as comparable, yet they remain the most commonly used, probably due to the much lower cost of small animal experimentation. One of the limitations of rodent models is the very large infarct size after LAD binding (60–80%). This is not representative of the clinical situation where infarcts tend to be smaller, usually about 10% of the left ventricle [12]. The rodent models do not allow for fine tuning the infarct size, due to anatomical and technical reasons. Very small changes in the location of the LAD binding site in the mouse model can lead to a shift from an extensive infarct to no infarct [20].

3.4.1.2 Cryoinjury

A model for myocardial damage as a result of very low temperatures has been proposed as an alternative to the LAD ligation, where infarct size can be controlled in a better fashion. A probe of varying size that is chilled down with liquid nitrogen and then applied to the exposed myocardium for a fixed amount of time resulting in cell death and focal necrosis [21, 22]. This method is technically easier than LAD ligation but from a physiological point of view does not reflect the pathological processes that occur during an ischemic event. Another limitation is that the heart failure that results from this damage is usually not overt and therefore the evaluation

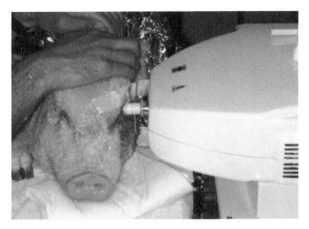

Figure 3.2   *In vivo* confocal microscopy is used after transplantation as a non-invasive means to track in-growth or overgrowth of cells and nerves into the implant, and also to detect any undesired issues, e.g. neovascularisation, imflammatory cell invasion.

models. Knock-out animals allow for the removal of any element of a response pathway and how this potentially affects the immunological outcome can be determined [8].

3.4.2.2 Lamellar and penetrating keratoplasty
Full-thickness replacement of the cornea is referred to as penetrating kerato-plasty (PK). Lamellar keratoplasty (LK) refers to partial-thickness corneal transplantation, which can reach all the way to Descemet's membrane, the layer just before the corneal endothelium. The preservation of the endothelial layer is significant for the outcome of grafting,

The larger animal models are appropriate for both lamellar and penetrating keratoplasty, however the rabbit reacts very strongly to any entry into the anterior chamber, leading to extensive clotting. This is a limiting factor in the case of PK [8].

3.4.2.3 Infectious models
In both humans and animals, there are a range of pathogens that infect the cornea and where the infection leads to severe immunopathological reactions, the cornea is scarred and requires transplantation with a donor graft. However, there is a severe shortage of human donor corneas worldwide and also in a number of conditions, donor grafting is contraindicated. An example is in the

case of Herpes Simplex Virus-1 (HSV-1) keratitis, where the virus remains latent within the host and can reactivate to cause active disease and rejection of the donor graft. Animal models of viral and bacterial infection have been established, most of them in rodents [27–29].

3.5 *In Vitro* Systems as Alternatives to Animal Testing

The demand to replace research on live animals when possible makes it necessary to explore alternative methods before commencing in vivo experiments. *In vitro* methods are faster and more cost effective compared to animal experimentation. Initial screening of new materials or compounds using cell lines to assess biocompatibility and to detect possible cytotoxicity is a standard. Organotypic culture is another alternative to animal trials, giving the benefit of a maintained tissue structure and sometimes functions.

An example of this is the beating heart slice culture system. Heart slices obtained from newborn rats sustain spontaneous beating for weeks to months if cultivated in the air-liquid interface [30]. This model has been used for testing cell engraftment when developing stem cell therapies for heart regeneration. The rhythmic beating of the heart slices is important to evaluate how the contractions of the heart tissue could affect materials or cells being introduced, as engraftment is less challenging in a static tissue. Whole organ cultures have also been developed, e.g. the isolated heart [31]

Within our group, we have developed 3-dimensional organotypic equivalents to the human cornea, which we have used in pre-screening all our biomaterials prior to testing within animals. This is detailed below.

Apart from being simply cost effective alternative to animal trials, organotypic cultures allow for studies of specific cell interactions with bioactive factors that would be possible in whole animals due to the presence of often confounding systemic effects.

3.5.1 *In Vitro* Corneal Equivalents for Screening Biomaterials as Potential Implants

In 1999, Griffith et al. reported the development of the first 3-dimensional human corneal equivalent made using human corneal cell lines that reproduced the key morphology and functional characteristics of the human cornea [32]. This was followed up by innervation of the whole system that allowed for

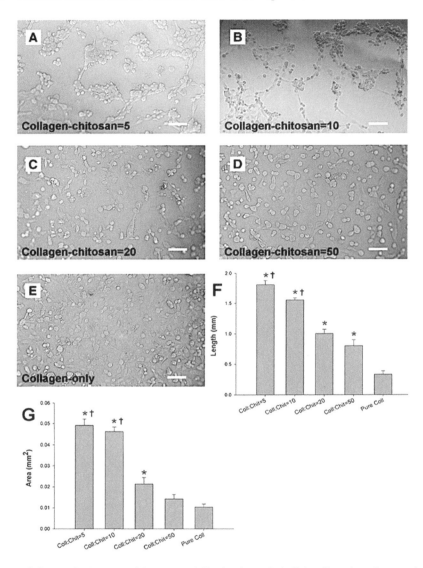

Figure 3.6   (A–E) Images of human umbilical vein endothelial cells cultured on various hydrogels after 2 days. Capillary-like networks were formed on (A) 5:1 and (B) 10:1 ratio collagen–chitosan hydrogels. (F) The average complete tube length on the different hydrogels. *p≤0.003 versus collagen; {p≤0.04 versus all other groups. Scale bar¼75 mm. (G) The average complete area of tubule structure formation on the different hydrogels. *p≤0.005 versus collagen; {p≤0.0004 versus all other groups. Reproduced from Deng C et al. (2010) [36].

[13] Sun Y. Myocardial repair/remodelling following infarction: Roles of local factors. Cardiovascular research 2009; 81:482–90.

[14] Hasenfuss G. Animal models of human cardiovascular disease, heart failure and hypertrophy. Cardiovascular research 1998; 39:60–76.

[15] Ytrehus K. The ischemic heart–experimental models. Pharmacological research: the official journal of the Italian Pharmacological Society 2000; 42:193–203.

[16] Arvola L, Bertelsen E, Lochner A, Ytrehus K. Sustained anti-beta-adrenergic effect of melatonin in guinea pig heart papillary muscle. Scandinavian cardiovascular journal: SCJ 2006; 40:37–42.

[17] Abarbanell AM, Herrmann JL, Weil BR, Wang Y, Tan J, Moberly SP, et al. Animal models of myocardial and vascular injury. The Journal of surgical research 2010; 162:239–49.

[18] Fröhlich GM, Meier P, White SK, Yellon DM, Hausenloy DJ. Myocardial reperfusion injury: Looking beyond primary pci. Eur Heart J 2013; 34:1714–22.

[19] Michael LH, Entman ML, Hartley CJ, Youker KA, Zhu J, Hall SR, et al. Myocardial ischemia and reperfusion: A murine model. The American journal of physiology 1995; 269:H2147–54.

[20] Degabriele NM, Griesenbach U, Sato K, Post MJ, Zhu J, Williams J, et al. Critical appraisal of the mouse model of myocardial infarction. Experimental physiology 2004; 89:497–505.

[21] van den Bos EJ, Mees BM, de Waard MC, de Crom R, Duncker DJ. A novel model of cryoinjury-induced myocardial infarction in the mouse: A comparison with coronary artery ligation. American journal of physiology Heart and circulatory physiology 2005; 289:H1291–300.

[22] van Amerongen MJ, Harmsen MC, Petersen AH, Popa ER, van Luyn MJ. Cryoinjury: A model of myocardial regeneration. Cardiovascular pathology: the official journal of the Society for Cardiovascular Pathology 2008; 17:23–31.

[23] Kuraitis D, Zhang P, Zhang Y, Padavan DT, McEwan K, Sofrenovic T, et al. A stromal cell-derived factor-1 releasing matrix enhances the progenitor cell response and blood vessel growth in ischaemic skeletal muscle. European cells & materials 2011; 22:109–23.

[24] Kuraitis D, Zhang P, McEwan K, Zhang J, McKee D, Sofrenovic T, et al. Controlled release of stromal cell-derived factor-1 for enhanced progenitor response in ischemia. Journal of controlled release: official journal of the Controlled Release Society 2011; 152 Suppl 1: e216–8.

[25] Li F, Carlsson D, Lohmann C, Suuronen E, Vascotto S, Kobuch K, et al. Cellular and nerve regeneration within a biosynthetic extracellular matrix for corneal transplantation. Proceedings of the National Academy of Sciences of the United States of America 2003; 100:15346–51.

[26] Liu W, Deng C, McLaughlin CR, Fagerholm P, Lagali NS, Heyne B, et al. Collagen-phosphorylcholine interpenetrating network hydrogels as corneal substitutes. Biomaterials 2009; 30:1551–9.

[27] Cowell BA, Wu C, Fleiszig SMA. Use of an animal model in studies of bacterial corneal infection. Inst Lab Animal Res J 1999; 40:43–50.

[28] Webre JM, Hill JM, Nolan NM, Clement C, McFerrin HE, Bhattacharjee PS, et al. Rabbit and mouse models of hsv-1 latency, reactivation, and recurrent eye diseases. Journal of biomedicine & biotechnology 2012; 2012:612316.

[29] Stuart PM, Keadle TL. Recurrent herpetic stromal keratitis in mice: A model for studying human hsk. Clinical & developmental immunology 2012; 2012:728480.

[30] Habeler W, Pouillot S, Plancheron A, Puceat M, Peschanski M, Monville C. An *in vitro* beating heart model for long-term assessment of experimental therapeutics. Cardiovascular research 2009; 81:253–9.

[31] Asimakis GK, Inners-McBride K, Medellin G, Conti VR. Ischemic preconditioning attenuates acidosis and postischemic dysfunction in isolated rat heart. The American journal of physiology 1992; 263:H887–94.

[32] Griffith M, Osborne R, Munger R, Xiong X, Doillon CJ, Laycock NL, et al. Functional human corneal equivalents constructed from cell lines. Science 1999; 286:2169–72.

[33] Suuronen EJ, McLaughlin CR, Stys PK, Nakamura M, Munger R, Griffith M. Functional innervation in tissue engineered models for *in vitro* study and testing purposes. Toxicological sciences: an official journal of the Society of Toxicology 2004; 82:525–33.

[34] Vailhe B, Vittet D, Feige JJ. *In vitro* models of vasculogenesis and angiogenesis. Laboratory investigation; a journal of technical methods and pathology 2001; 81:439–52.

[35] Nakatsu MN, Hughes CC. An optimized three-dimensional *in vitro* model for the analysis of angiogenesis. Methods in enzymology 2008; 443:65–82.

[36] Deng C, Zhang P, Vulesevic B, Kuraitis D, Li F, Yang AF, et al. A collagen-chitosan hydrogel for endothelial differentiation and angiogenesis. Tissue engineering Part A 2010; 16:3099–109.

4.1 Introduction

Most of the evidence that pluripotent stem cells can be directed to differentiate into specific types of cells suitable for transplantation comes from experiments with mouse cells, and offers the cues for translational research. The application of stem cells in human regenerative medicine could be an alternative to organ transplantation, avoiding the problem of donor shortage and rejection [1]. One important question that has to be answered before stem cells can be effectively translated into significant medical treatments is what type of pluripotent stem cells are the most suitable for human clinical application.

Embryonic stem cells (ESC) are the most versatile cells among pluripotent stem cells. These cells, derived from the inner cells mass of blastocysts, are able to give rise to all type of adult differentiated cells. The development of human embryonic stem cells (hESC) gave an incredible acceleration to stem cell research [2, 3]. Human ESC, like murine ESC, can be differentiated into tissue derived from all three germ layers and have a limitless reproductive capacity. Despite their huge potentiality, the use of human ESC in cell therapy is impeded by moral and ethical concern of destroying human embryos for derivation of ESCs [4]. The recent discovery of the ability to reprogram adult cells into pluripotent embryonic-like stem cells (known as induced pluripotent stem cells; iPS) has profound implications for stem cell therapy [5–8]. The first generation of iPS was generated by introduction of transcription factors, including c-Myc, by retroviral vectors, this probably lead to the generation of neoplastic cells from some induced cells. This problem was solved by using alternative vectors that do not comprise c-Myc [9]. Rapid progress has been made in finding alternative ways to reprogram cells. Now virus-free iPS are available from adult somatic cells [10, 11], this could have important implication in terms of clinical application. iPS have shown remarkable promises in many ways, including the generation of patient-specific iPS [12]. However, the drawback of iPS-based therapy is the need of transducing cells with reprogramming factors to achieve an efficient generation of iPS. Moreover, the existence of inherent epigenetic differences between iPS and ESC can affect iPS functionality [13, 14]. Adult stem cells have been identified in several human tissues, such as liver [15], blood [16], skin [17] and testis [18]. Adult stem cells could be an autologous, free-from-ethical concern, source of pluripotent stem cells. A particular type of adult stem cells that have attracted the interest of the scientific community in the last years are spermatogonial stem cells. Spermatogonial stem cells (SSCs) reside in the basal membrane of testis, these cells are the only stem

4.2 Hepatocytes Derived from GPSCs

Hepatic disorders affect hundreds of millions of people worldwide. The mild conditions, if not cured properly, may lead to progressive liver injury, liver fibrosis and ultimately cirrhosis, portal hypertension, liver failure, and, in some instances, cancer [38]. Orthotopic liver transplantation (OLT) has become the standard of care for the treatment of patients with end-stage liver disease resulting in elevated request for OLT. However, the ongoing organ shortage has impeded the treatment of all recurrent-incurable hepatic diseases. Importantly, adult hepatocytes are capable of replicating under particular conditions, for e.g. after a partial hepatectomy [39]. The human liver also contains resident stem cells and the bipotential oval cells that can differentiate into hepatocytes when necessary [15, 40]. However, in a severely compromised liver, the regenerative capacity of hepatocytes and liver stem cells is impaired and can no longer restore functionality. Thus, cell therapy may be the only way out. Hepatocytes transplant have been reported in several cases. For e.g., repeated hepatocyte transplantation in a patient with acute liver failure due to mushroom poisoning has been shown to improve patient's condition and liver functionality [41]. The reported viability of the thawed primary hepatocytes was 62%. This loss in cell viability upon thawing may be a problem, especially when billions of cells are needed for transplant in patients. Hence, alternative sources of hepatocytes are being looked for. Human bone marrow-derived stem cells have been shown to differentiate to hepatocytes *in vitro* and reverse liver failure *in vivo*, but these cells are present in minute fractions and thus are tedious to isolate and difficult to expand [42]. ESCs are considered a very promising source of hepatocytes for cell therapy due to their limitless capacity for self-renewal and proliferation, and their ability to differentiate into all major cell lineages [42]. However, the allogenic nature of these cells as well as the ethical burden, has impeded their use in the clinical setting. GPSCs are an interesting alternative. The differentiative potential of GPSCs are being extensively studied in the mouse with the aim of extending the results to human. EBs generated from mouse GPSCs express the early hepatic marker, alpha fetoprotein. We and others have reported on the directed differentiation of GPSCs into hepatocyte-like cells [36, 43]. Metabolically active hepatocytes, capable of albumin and haptoglobin secretion, urea synthesis, glycogen storage, and indocyanine green uptake can be derived from GPSCs *in vitro* [36]. Our large scale microarray analysis comparing GPSCs to ESCs during hepatocyte differentiation revealed that there was a marked similarity in gene expression profile between these two cell lines. The GPSC-derived hepatocytes, at Day 28 of *in vitro* differentiation, were closer to fetal hepatocytes (embryonic day 16) than adult hepatocytes (post natal day 1) [36]. *In vivo* studies in mouse models of liver diseases will reveal if these GPSC-derived hepatocytes can home to liver and restore functionality.

GPSCs are thus a promising tool for the treatment of liver diseases. Hepatocytes derived from GPSCs may provide the minimal amount of functional protein that is required to correct certain metabolic diseases. These cells may also be useful in the period awaiting transplant as in the case of newborns with genetic defects [44]. As liver transplant requires mainly blood group compatibility, there is the possibility of transplanting GPSC-derived hepatocytes in both male and female patients [45].

4.3 Cardiac Cells Derived from GPSCs

Heart transplantation is one the most effective treatment for severe cardiomyopathies. However, the insufficient number of matching donor hearts has elicited search for other sources of cardiomyocytes including extracardiac ones. Transplantable cell sources identified hitherto comprise cardiac progenitor cells [46], mesenchymal stem cells [47], fetal cardiomyocytes [48], bone marrow cells [49], ES cells [50] and iPS cells [51]. Due to the advantages described above, GPSCs may also offer a substitute to ES cells or iPS cells in cardiomyocytes generation. EBs formed from mouse GPSCs form contracting areas which show cardiomyocyte phenotype characterized by sarcomeric striations when stained for α-sarcomeric actinin, sarcomeric MHC and cardiac troponin T, organized in bundles [32]. Functional cardiomyocytes could also be generated from these GPSC-derived EBs [34]. Through molecular, cellular, and physiological assays, Guan et al. demonstrated that GPSC-derived cardiomyocytes engrafted into the left ventricular free wall of mice one month post-injection. These cells proliferated in the normal heart without giving rise to teratoma. However, no spontaneous *in vivo* differentiation into cardiomyocytes were observed in the normal heart. The GPSCs differentiated *in loco* into vascular endothelial and smooth muscle cells probably related to the fact that these cells were not induced to differentiate into cardiomycoytes prior to transplantation [52]. Interestingly, Flk1+ cells from differentiating GPSCs could give rise to mature cardiomyocytes and endothelial cells as efficiently as ES cells [53]. Transplantation of these Flk1+ GPSCs directly into the heart of ischemic mice improved cardiac function [54]. Four weeks after treatment,

References

[1] Chistiakov DA. Liver Regenerative Medicine: Advances and Challenges. *Cells Tissues Organs*. 2012.

[2] Thomson JA, Itskovitz-Eldor J, Shapiro SS, Waknitz MA, Swiergiel JJ, Marshall VS, Jones JM. Embryonic stem cell lines derived from human blastocysts. *Science*. 1998; 282(5391): 1145–1147.

[3] Guhr A, Kurtz A, Friedgen K, Loser P. Current state of human embryonic stem cell research: an overview of cell lines and their use in experimental work. *Stem Cells*. 2006; 24(10): 2187–2191.

[4] de Wert G, Mummery C. Human embryonic stem cells: research, ethics and policy. *Hum Reprod*. 2003; 18(4): 672–682.

[5] Takahashi K, Yamanaka S. Induction of pluripotent stem cells from mouse embryonic and adult fibroblast cultures by defined factors. *Cell*. 2006; 126(4): 663–676

[6] Gibson SA, Gao GD, McDonagh K, Shen S. Progress on stem cell research towards the treatment of Parkinson's disease. *Stem Cell Res Ther*. 2012; 3(2): 11

[7] Takahashi K, Tanabe K, Ohnuki M, Narita M, Ichisaka T, Tomoda K, Yamanaka S. Induction of pluripotent stem cells from adult human fibroblasts by defined factors. *Cell*. 2007; 131(5): 861–872.

[8] Yu J, Vodyanik MA, Smuga-Otto K, Antosiewicz-Bourget J, Frane JL, Tian S, Nie J, Jonsdottir GA, Ruotti V, Stewart R, Slukvin, II, Thomson JA. Induced pluripotent stem cell lines derived from human somatic cells. *Science*. 2007; 318(5858): 1917–1920.

[9] Nakagawa M, Koyanagi M, Tanabe K, Takahashi K, Ichisaka T, Aoi T, Okita K, Mochiduki Y, Takizawa N, Yamanaka S. Generation of induced pluripotent stem cells without Myc from mouse and human fibroblasts. *Nat Biotechnol*. 2008; 26(1): 101–106.

[10] Kaji K, Norrby K, Paca A, Mileikovsky M, Mohseni P, Woltjen K. Virus-free induction of pluripotency and subsequent excision of reprogramming factors. *Nature*. 2009; 458(7239): 771–775.

[11] Lengner CJ. iPS cell technology in regenerative medicine. *Ann N Y Acad Sci*. 2010; 1192: 38–44.

[12] Dimos JT, Rodolfa KT, Niakan KK, Weisenthal LM, Mitsumoto H, Chung W, Croft GF, Saphier G, Leibel R, Goland R, Wichterle H, Henderson CE, Eggan K. Induced pluripotent stem cells generated from patients with ALS can be differentiated into motor neurons. *Science*. 2008; 321(5893): 1218–1221.

stem cells from cultured human primordial germ cells. *Proc Natl Acad Sci U S A*. 1998; 95(23): 13726–13731.

[27] Kanatsu-Shinohara M, Inoue K, Lee J, Yoshimoto M, Ogonuki N, Miki H, Baba S, Kato T, Kazuki Y, Toyokuni S, Toyoshima M, Niwa O, Oshimura M, Heike T, Nakahata T, Ishino F, Ogura A, Shinohara T. Generation of pluripotent stem cells from neonatal mouse testis. *Cell*. 2004; 119(7): 1001–1012.

[28] Seandel M, James D, Shmelkov SV, Falciatori I, Kim J, Chavala S, Scherr DS, Zhang F, Torres R, Gale NW, Yancopoulos GD, Murphy A, Valenzuela DM, Hobbs RM, Pandolfi PP, Rafii S. Generation of functional multipotent adult stem cells from GPR125+ germline progenitors. *Nature*. 2007; 449(7160): 346–350.

[29] Conrad S, Renninger M, Hennenlotter J, Wiesner T, Just L, Bonin M, Aicher W, Buhring HJ, Mattheus U, Mack A, Wagner HJ, Minger S, Matzkies M, Reppel M, Hescheler J, Sievert KD, Stenzl A, Skutella T. Generation of pluripotent stem cells from adult human testis. *Nature*. 2008; 456(7220): 344–349.

[30] Golestaneh N, Kokkinaki M, Pant D, Jiang J, DeStefano D, Fernandez-Bueno C, Rone JD, Haddad BR, Gallicano GI, Dym M. Pluripotent stem cells derived from adult human testes. *Stem Cells Dev*. 2009; 18(8): 1115–1126.

[31] Kossack N, Meneses J, Shefi S, Nguyen HN, Chavez S, Nicholas C, Gromoll J, Turek PJ, Reijo-Pera RA. Isolation and characterization of pluripotent human spermatogonial stem cell-derived cells. *Stem Cells*. 2009; 27(1): 138–149.

[32] Guan K, Nayernia K, Maier LS, Wagner S, Dressel R, Lee JH, Nolte J, Wolf F, Li M, Engel W, Hasenfuss G. Pluripotency of spermatogonial stem cells from adult mouse testis. *Nature*. 2006; 440(7088): 1199–1203.

[33] Mizrak SC, Chikhovskaya JV, Sadri-Ardekani H, van Daalen S, Korver CM, Hovingh SE, Roepers-Gajadien HL, Raya A, Fluiter K, de Reijke TM, de la Rosette JJ, Knegt AC, Belmonte JC, van der Veen F, de Rooij DG, Repping S, van Pelt AM. Embryonic stem cell-like cells derived from adult human testis. *Hum Reprod*. 2010; 25(1): 158–167.

[34] Guan K, Wagner S, Unsold B, Maier LS, Kaiser D, Hemmerlein B, Nayernia K, Engel W, Hasenfuss G. Generation of functional cardiomyocytes from adult mouse spermatogonial stem cells. *Circ Res*. 2007; 100(11): 1615–1625.

[35] Streckfuss-Bomeke K, Vlasov A, Hulsmann S, Yin D, Nayernia K, Engel W, Hasenfuss G, Guan K. Generation of functional neurons and glia from multipotent adult mouse germ-line stem cells. *Stem Cell Res*. 2009; 2(2): 139–154.

[36] Fagoonee S, Hobbs RM, De Chiara L, Cantarella D, Piro RM, Tolosano E, Medico E, Provero P, Pandolfi PP, Silengo L, Altruda F. Generation of functional hepatocytes from mouse germ line cell-derived pluripotent stem cells *in vitro*. *Stem Cells Dev*. 2010; 19(8): 1183–1194.

[37] Chikhovskaya JV, Jonker MJ, Meissner A, Breit TM, Repping S, van Pelt AM. Human testis-derived embryonic stem cell-like cells are not pluripotent, but possess potential of mesenchymal progenitors. *Hum Reprod*. 2012; 27(1): 210–221.

[38] Piscaglia AC, Campanale M, Gasbarrini A, Gasbarrini G. Stem cell-based therapies for liver diseases: state of the art and new perspectives. *Stem Cells Int*. 2010; 2010: 259–461.

[39] Brezillon N, Kremsdorf D, Weiss MC. Cell therapy for the diseased liver: from stem cell biology to novel models for hepatotropic human pathogens. *Dis Model Mech*. 2008; 1(2–3): 113–130.

[40] Oh SH, Hatch HM, Petersen BE. Hepatic oval 'stem' cell in liver regeneration. *Semin Cell Dev Biol*. 2002; 13(6): 405–409.

[41] Schneider A, Attaran M, Meier PN, Strassburg C, Manns MP, Ott M, Barthold M, Arseniev L, Becker T, Panning B. Hepatocyte transplantation in an acute liver failure due to mushroom poisoning. *Transplantation*. 2006; 82(8): 1115–1116.

[42] Cho CH, Parashurama N, Park EY, Suganuma K, Nahmias Y, Park J, Tilles AW, Berthiaume F, Yarmush ML. Homogeneous differentiation of hepatocyte-like cells from embryonic stem cells: applications for the treatment of liver failure. *FASEB J*. 2008; 22(3): 898–909.

[43] Loya K, Eggenschwiler R, Ko K, Sgodda M, Andre F, Bleidissel M, Scholer HR, Cantz T. Hepatic differentiation of pluripotent stem cells. *Biol Chem*. 2009; 390(10): 1047–1055.

[44] Cowles RA, Lobritto SJ, Ventura KA, Harren PA, Gelbard R, Emond JC, Altman RP, Jan DM. Timing of liver transplantation in biliary atresia-results in 71 children managed by a multidisciplinary team. *J Pediatr Surg*. 2008; 43(9): 1605–1609.

[45] Navarro V, Herrine S, Katopes C, Colombe B, Spain CV. The effect of HLA class I (A and B) and class II (DR) compatibility on liver transplantation outcomes: an analysis of the OPTN database. *Liver Transpl*. 2006; 12(4): 652–658.

crankshaft. Common to these cases is the idea that the material is inert, i.e. does not change its properties except from the damage caused by the applied loads.

Stress and strain are central concepts in strength of materials, and the concept of strength is coupled to an idea of the local stresses or strains causing immediate or cumulative damage at each material point. For materials with linear behavior (which is usually the case for sufficiently small deformations) stresses are coupled to strains through Hooke's law, stating that

$$\sigma = \mathbf{E}\varepsilon \tag{5.1}$$

where σ is the stress tensor, ε is the strain tensor, and $\mathbf{E}$ is the elasticity tensor. Two points are important for later application of this idea to living tissues. Firstly, stress cannot be measured but only calculated, whereas strain is derived directly from deformation, which can be experimentally observed. It may not be immediately obvious to the casual student of mechanics that force (and stress) are imaginary physical quantities, but it becomes obvious if we look at methods for measuring force: they are all based on some form of observation of deformation.

Secondly, all terms of Equation (5.1) are tensors, i.e. multidimensional properties. In its most general form, $\mathbf{E}$ contains 81 components, which by means of symmetry conditions and thermodynamic considerations can be drastically reduced. All properties, however, remain multidimensional to some extent depending on the properties of the material in question. It therefore rarely makes sense to discuss "the strain" or "the stress" as though the property is one-dimensional.

It is therefore also impossible, except in much idealized cases, to appoint a single stress or strain level that breaks a given material. However, many technically important materials can with good approximation be assumed ductile, homogeneous and isotropic and, for such materials, the yield criterion of von Mises is well-established. This criterion is based on a scalar combination of components of the stress tensor. We shall call this scalar combination the von Mises stress. For a certain class of materials, it is well accepted that yield, i.e. permanent deformation, occurs when the von Mises stress exceeds a magnitude that is characteristic for the given material. It is important to notice that many different combinations of stress components can result in the same value of the von Mises stress and therefore lead to yield. Similar yield or failure criteria have been developed with varying success for more complex materials, for instance the Tsai-Hill [2] and Tsai-Wu [3] criteria for composites or statistical failure prediction based on Weibull statistics [4] for ceramics.

Let us review the definition of the von Mises stress. A material has six stress components in its tensor. These components depend on the coordinate system, so the stress tensor changes when the reference frame is rotated, although the material remains in the same state. In other words, the same stress state in the same material point can be expressed by many different stress tensors. In its general form, a stress tensor contains the following components:

$$\sigma = \begin{bmatrix} \sigma_{xx} & \sigma_{xy} & \sigma_{xz} \\ \sigma_{xy} & \sigma_{yy} & \sigma_{yz} \\ \sigma_{xz} & \sigma_{yz} & \sigma_{zz} \end{bmatrix} \qquad (5.2)$$

where indices x, y and z refer to directions in the chosen coordinate system. The diagonal elements, σ_{ii}, are normal stresses and designate pure compression or tension of the material, and the off-diagonal elements are shear stresses. The tensor is always symmetrical and positive definite for any physical material.

It turns out that stress tensors have the interesting property that it is always possible to rotate the coordinate system such that the tensor takes the simplified form:

$$\sigma = \begin{bmatrix} \sigma_1 & 0 & 0 \\ 0 & \sigma_2 & 0 \\ 0 & 0 & \sigma_3 \end{bmatrix} \qquad (5.3)$$

In this particular rotation of the coordinate system, the material experiences no shear stress. The normal stresses in this state, σ_1, σ_2 and σ_3, are called principal stresses, and $\sigma_1 \geq \sigma_2 \geq \sigma_3$ by definition. Most people have an immediate physical comprehension of the difference between shear and normal deformation and are able to recognize these states when they see them applied to a soft material with sufficiently large deformations. It is mind-boggling that viewing the same material point in the same state of deformation from a different vantage point would reveal no shear. That is, however, a mathematical fact.

There exists another rotation of the coordinate system where the shear stresses are at their maximum, and it furthermore turns out that these shear stresses are equivalent to the so-called deviations of the principal stresses: $\sigma_1 - \sigma_3$, $\sigma_1 - \sigma_2$, and $\sigma_2 - \sigma_3$. We can therefore conclude that there exists only a single state of stress that has no shear, namely the case where $\sigma_1 = \sigma_2 = \sigma_3$. This case of similar stress in all directions is called a hydrostatic

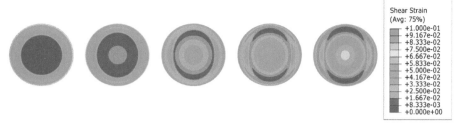

Figure 5.2   Finite element simulation of maximum shear strains on a membrane draped over posts with varying degrees of ovality.

While it is theoretically possible to control the amount of shear strain imposed on a tissue sample by means of non-circular posts, it unfortunately turns out that the adhesion between the silicon membrane and the tissue sample is insufficient to cause tissue injury. When imposing strain levels similar to those experienced by, for instance, muscle tissue in the buttocks in the seated posture, the cells come loose from the membrane and cease to follow the membrane's deformation. Thus, this method is not suitable for investigations of tissue injury, but it does allow for mechanical stimulation of tissue samples for other purposes [6].

5.2.1 Hertz-inspired Tissue Deformation

An alternative loading mechanism, inspired by Hertz contact mechanics [7], was invented to enable imposition of sufficient strain to cause necrosis. Mechanical problems of contact are in general highly nonlinear and very challenging, but a famous analytical solution attributed to Heinrich Hertz covers the special case of two linearly elastic spheres in contact. Given the material properties, the radii of the two spheres and the compression force, the analytical solution predicts the deformation state including strain and stress fields in the two parts. This is one of the truly classical problems of mechanics and it is an important part of the basis of the field of tribology and the ability to develop many important machine parts such as bearings and gears. An important special case of Hertz' solution is when one of the spheres has an infinite radius, i.e. is flat.

The case of a sphere pressed into a planar surface is axisymmetric and so is the resulting strain state. If we focus on the planar part, this means that the strain tensor in a given point depends only on the force (or relative displacement of the two parts), the point's depth under the planar surface

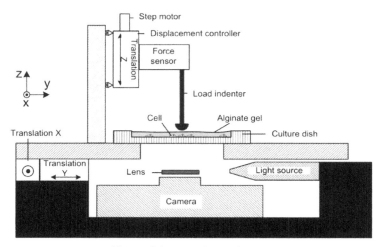

Figure 5.4   Experimental setup.

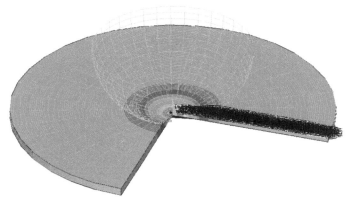

Figure 5.5   Axisymmetric, nonlinear finite element model of sphere indentation into an alginate gel.

The finite element model predicts the strain field under the indenter and consequently also the strain state felt by a cell located at a given point in the gel. It is possible to visualize the strain field in terms of maximum shear strain to obtain the strain map of Figure 5.6.

The fact that the strain field decreases systematically (albeit nonlinearly) from the center of pressure to zero at about 8 mm distance means that the viability of cells under varying shear strain can be studied systematically using this setup.

Figure 5.6 Maximum shear strain as a function of radius (r) and height (z) in a gel under compression of a circular indenter. The strain is at its maximum at (r, z) = (0, 0) i.e. directly below the center of pressure and decreases radially.

The viability of the cells is observed under microscope by means of Lentiviral transfection of emerald green fluorescent protein (GFP). This staining technique allows the simultaneous study of cell morphology and viability by combination of the red and green channels of the microscope image. Necrotic cells appear as bright spots in the image.

5.2.2 Preliminary Results of Cell Straining

Figure 5.7 shows the viability of cells under the indenter, where the yellow dots designate dead cells. In this test, the indentation was increased over time to compensate for a significant stress relaxation of the gel.

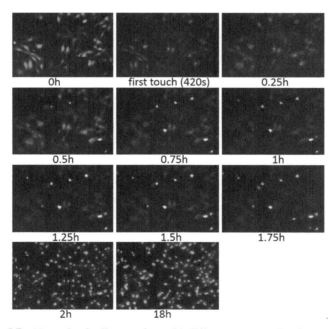

Figure 5.7 Necrosis of cells over time with different compression forces applied.

neural cells. We explain the over-arching chapter role of the immune properties of MSCs in the translation of MSCs as well as safety issues for clinical application.

Keywords: mesenchymal stem cells, immunosuppression, bone marrow, cytokine, cancer, stem cells.

6.1 Introduction

Stem cells have the potential to self-renew and differentiate into all cell types, making stem cells as the future therapy for tissue regeneration and organogenesis. Embryonic stem cells are linked to ethical concerns and scientifically, in tumor formation. These issues dampen the regenerative potential of embryonic stem cells. Similar arguments can be also made for induced pluripotent stem cells. Thus, clinical application is narrowed to adult stem cells such as those in the brain and in bone marrow, such as mesenchymal stem cells (MSCs). The clinical and experimental data indicated that MSCs can have regenerative/repair potential for several clinical disorders (www.clinicaltrials.gov).

The multipotent mesenchymal stem cells are primordial in origin and can be isolated from fetal and adult tissues such as the placenta, bone marrow and adipose tissues [1–4]. Several membrane proteins have been identified to select and to phenotype MSCs. These include, but not limited to CD73, CD90, CD105. MSCs do not express markers like CD45, CD34, CD14, CD19. In addition to phenotype, other molecules have been proposed as methods to identify MSCs. These include vimentin and fibronectin. In all studies, the function of MSCs need to be validated for multipotency. In general, it is acceptable to induce MSCs to differentiate into osteoblast, adipocyte and chondrocyte [5].

MSCs are attractive for stem cell treatment, mostly due to reduced ethical concerns and ease of *in vitro* expansion. Furthermore, there is no need for a match at the major histocompatibility complex; thus making MSCs available as 'off the shelf' sources in cell therapy. The microenvironment plays an important role in the functional response of MSCs. The immune function of MSCs is particularly relevant. MSCs can be immune enhancer and suppressor cells. The immune function of MSCs depends on the milieu of the microenvironment. Specific cytokines and chemokines can be chemoattractant to facilitate the migration and homing of MSCs and other immune cells to the site of tissue injury.

6.2 MSC Immunology

MSCs show functional plasticity with regards to their immune properties by exerting both immune suppressor and enhancer functions [6], producing various cytokines that can stimulate the cells through autocrine and/or paracrine manner [7]. It has been suggested that major histocompatibility complex-II (MHC-II) expression be included among the minimum requirements for designating a cell as MSC [5]. However, there are several reports of MHC-II-negative cells with a phenotype and multi-lineage capacity similar to MSCs [8], suggesting that a population of MSCs do not express (detectable amounts of) MHC-II.

MHC-II allows cells to act as antigen presenting cells (APCs), with the cell mounting the antigen within the MHC-II grove to activate CD4+ T cells. The MSCs expressing MHC-II may then be able to act as APCs. However, unlike most APCs, MSCs express MHC-II in a bimodal fashion, with high MHC-II densities at low levels of interferon gamma (IFNγ), and low MHC-II density at high IFNγ levels [9]. This is highly significant when considering MSCs as a therapeutic tool, as the MSCs would be in an inflammatory microenvironment, in the presence of a milieu of inflammatory cytokines – including IFNγ. This bimodal activity has been observed *in vitro* in MSC-derived neurons, whereby the neurons expressed low levels of MHC-II, but MHC-II level could be restored by stimulation with IFNγ [10]. Thus MHC-II expression has the potential to be problematic in case the MHC-II is re-expressed. If so, this could result in rejection of the transplanted MSCs.

Since the majority of MSCs show low to undetectable expression of MHC-II molecules on their cell surface they fail to activate T-cell response [5, 8]. In the absence of an allogeneic response, MSCs are suitable candidates for allogeneic transplant [6]. In the absence of pro- and anti-inflammatory cytokines, MSCs are further immunoprivileged in that they interfere with other immune cell functions, such as inhibiting B-cell proliferation and chemotaxis [11], suppressing the activity of dendritic and natural killer cells [12], and triggering the proliferation of regulatory T-cells to suppress an immune response [13]. It has been suggested that these immunosuppressive properties may play a role in tissue repair, modulating the other immune cells to prevent immune inflicted tissue injury and promote healing [14]. An equilibrium of these two properties must be understood and balanced for normal MSC function, as well as for MSC function in regeneration and repair [6, 15].

An increase in the interest of the mechanisms of homing and migration of MSCs to tumors, has led to a better understanding of this process. There are several different molecules that have been found to be involved. Although the candidate molecules vary with the cancer type, they include growth factors, chemokines, and cytokines that are released from the tumor or surrounding stroma. One example is stromal cell derived factor 1α (SDF-1α) and its receptor chemokine (C-X-C motif) receptor 4 (CXCR4) commonly expressed on cancer cells. MSCs have been found to use SDF-1α-CXCR4 signaling for migration to areas of inflammation, which is often common in the tumor microevironment [34].

Other factors, such as VEGF, can enhance tumor tropism of MSCs to tumors. Breast cancer and gliomas have been reported to express high levels of VEGF, which induces the migration and invasion of MSCs to tumors [35]. MSC migration may be also increased in response to irradiation and hypoxia. Radiation may lead to increased expression of inflammatory mediators to enhance the migration of MSC to the tumor [35]. Hypoxia is often associated with tumor progression and can lead to the production of IL-6 which acts in a paracrine fashion on MSCs, causing increased migration to the tumor [36]. The mechanism of MSC homing and migration to the tumor site continues to be elucidated. However, the clinical and experimental evidence provided information on the tropism of MSC to brain tumors.

One of the earlier reports showing MSCs migrating to the region of gliomas was indicated with an experimental model using rats [37]. Autologous MSCs were intracranially implanted into rats that developed gliomas. The MSCs migrated and dispersed within the tumor mass [37]. Subsequent studies with immunocompromised mice showed human MSCs migrating to the region of human gliomas [38]. The MSCs were injected into the ipsilateral and contralateral carotid arteries of the mice [38]. In other studies, rat MSCs were injected intratumorally and this resulted in the migration to the invasive rat glioma and to the distant tumor microsatellites [39]. The investigators also observed that the implanted MSCs avoided the normal brain gray matter [40].

Based on the above findings, MSCs show promise as a delivery system for toxic substances to the tumor while being able to avoid adverse effects of the drug on normal brain tissue. Given the dire need for improved therapy for GBM researchers have also started looking at ways to increase the efficacy of current treatments by sensitizing the cells. Our group has recently published on the chemosensitization of GBM cells through the transfer of functional anti-miR-9 within MSCs, by packaging it within the exosomes [41]. As a cellular vehicle, MSCs could deliver chemosensitizing reagents as adjuvant to other treatments.

6.4 Regenarative Potential

MSC was discovered by Friedenstein *et al.* and was referred as CFU-F (colony forming unit-fibroblasts). Similar cells were isolated from the bone marrow and with similar formation of colonies [46, 47]. MSCs have several functions, including support of hematopoiesis during transplantation with simultaneous decrease of graft versus host through veto property [48].

MSCs have been shown to restore heart function [49, 50]. At a high dose of MSC when injected into the intracoronary region of left ventricle showed significant improvement in the normal function of the heart [49, 50]. This is a highly significant property of MSCs because cardiac failure is the leading cause of death in USA. The use of MSCs for neurological disorder has not yet reached the patients. However, the use of MSCs for neural disorder is widely accepted. Several animal models have shown full recovery of the damaged neurons. There are three possible explanations why MSCs might be important for neural repair. MSCs can differentiate into neurons, undergo cell fusion, release neurotropic factors to maintain the survival of the neurons and/or the release of non-neurotropic factors to promote the tissue repair [51]. Despite MSC is able to introduce repair and regeneration in the brain, it is still unclear if it is able to cross the blood brain barrier [52].

Human MSCs injected at the site of brain injury release neurotropic factors to induce endogenous recovery of damaged neurons [53]. This occurred by reduced inflammation, inhibition of apoptosis and increased proliferation and differentiation of neural stem cells. An early set of studies [54, 55] used a co-culture technique to shown that brain derived neurotropic factor (BDNF), glial derived neurotropic factor secreted by MSCs, induced neurite formation in neuroblastoma cell line. The role of BDNF was demonstrated with neutralizing antibodies, which prevented the regenerative potential of MSCs [54, 55]. In a spinal cord injury model with zebrafish, *TAC1* expression in MSCs improved the sensory and locomotors recovery by releasing some neurotropic factor [56]. Other examples of neurotropic factors can be nerve growth factor, neurotropin-3, ciliary neurotropic factor and vascular endothelial factor. In addition to neurotropic factors, *in vitro* studies showed that the extracellular matrix from MSCs can have a positive effect on the adhesion of neuronal cells by inducing neurite growth and astrocyte proliferation [57].

Both *in vitro* and *in vivo* studies suggest that MSCs can transdifferentiate into neurons, thereby providing these cells with the potential to promote neuronal repair. MSCs can be induced with defined condition and with cytokines to differentiate into neurons [56]. There are occasions if transdifferentiation of

MSC to neural cells can occur within a few hours and, whether the formation of neurons can be solely dependant on morphology [58, 59]. Subsequent studies indicated that transdifferentiation could happen only under stress and not under normal conditions [58–61]. In 2001, noggin, which can induce neural formation by inhibiting BMP2/4, TGF-ß, was used to induce neuron formation by transfecting MSCs. The noggin-transfectants expressed neuronal and astrocyte markers [62]. Retinoic acid (RA) was identified as an inducer of MSCs to form neurons with the expression of the neurotransmitter gene, *TAC1* with synaptic transmission [63]. There are several induction factors that can up regulate transdiffentiation of MSCs into neural cells such as matrigel [64]. The logic behind transdifferentiation of MSCs is not based solely on *in vitro* studies, as the conditions provided are totally artificial. When MSCs were injected to rat it migrated to the brain in the way neural stem cell (astrocyte engraft) migrated further losing MSC marker [65]. This showed the engrafted MSCs was well supported by the microenvironment of the rodents. The gradual increase in the neural marker in these MSCs showed its ability to transdifferentiate.

In Parkinson's disease (PD) rat model when hMSCs were injected it improved the motor function [66]. This was followed by clinical trial of using hMSCs in PD patient and it eventually lead to improvement in motor function with no significant side effect [67]. Amyotrophic Lateral Sclerosis (ALS) is a neurodegenerative disease caused by the death of motor neurons in cerebral cortex, brain stem and spinal cord. Human MSCs when transplanted in transgenic ALS mouse model with spinal cord injury improved the motor activity [68].

Patients with stroke who were injected with autologous MSCs showed transient improvement [69]. A subsequent repeat of the trial indicated that the transdifferentiated MSCs were able to retain their function for a prolonged period [69]. A recent *in vitro* study generated neurosphere-like aggregates from MSCs and then when injected them into a rat model of ischemic stroke to show that this method induced neuroprotection [70].

Other than neurotropic factor and transdifferentiaton, it is possible for the injected MSCs might fuse with the neural cells at the site of injury to cause functional improvement. Although evidence for this mechanism is limited there was a report in which MSCs were introduced into the rodent model [71, 72]. There was fusion with the neural cells and epigenetic changes with complete recovery of neuronal function [71, 72].

MSCs can be obtained and expanded easily without any ethical issues. These cells can differentiate into both mesenchymal and non-mesenchymal

[2] He, Q., C. Wan, and G. Li, Concise review: multipotent mesenchymal stromal cells in blood. Stem Cells, 2007. **25**(1): pp. 69–77.

[3] Lee, O. K., et al., Isolation of multipotent mesenchymal stem cells from umbilical cord blood. Blood, 2004. **103**(5): pp. 1669–75.

[4] Tsuda, H., et al., Allogenic fetal membrane-derived mesenchymal stem cells contribute to renal repair in experimental glomerulonephritis. Am J Physiol Renal Physiol, 2010. **299**(5): pp. F1004–13.

[5] Dominici, M., et al., Minimal criteria for defining multipotent mesenchymal stromal cells. The International Society for Cellular Therapy position statement. Cytotherapy, 2006. **8**(4): pp. 315–7.

[6] Sherman, L. S., et al., Moving from the laboratory bench to patients' bedside: considerations for effective therapy with stem cells. Clin Transl Sci, 2011. **4**(5): pp. 380–6.

[7] Castillo, M., et al., The immune properties of mesenchymal stem cells. Int J Biomed Sci, 2007. **3**(2): pp. 76–80.

[8] Jacobs, S. A., et al., Immunological characteristics of human mesenchymal stem cells and multipotent adult progenitor cells. Immunol Cell Biol, 2013. **91**(1): pp. 32–9.

[9] Tang, K. C., et al., Down-regulation of MHC II in mesenchymal stem cells at high IFN-gamma can be partly explained by cytoplasmic retention of CIITA. J Immunol, 2008. **180**(3): pp. 1826–33.

[10] Cheng, Z., et al., Targeted migration of mesenchymal stem cells modified with CXCR4 gene to infarcted myocardium improves cardiac performance. Mol Ther, 2008. **16**(3): pp. 571–9.

[11] Corcione, A., et al., Human mesenchymal stem cells modulate B-cell functions. Blood, 2006. **107**(1): pp. 367–72.

[12] De Miguel, M. P., et al., Immunosuppressive properties of mesenchymal stem cells: advances and applications. Curr Mol Med, 2012. **12**(5): pp. 574–91.

[13] Maccario, R., et al., Interaction of human mesenchymal stem cells with cells involved in alloantigen-specific immune response favors the differentiation of CD4+ T-cell subsets expressing a regulatory/suppressive phenotype. Haematologica, 2005. **90**(4): pp. 516–25.

[14] Hoogduijn, M. J., et al., The immunomodulatory properties of mesenchymal stem cells and their use for immunotherapy. Int Immunopharmacol, 2010. **10**(12): pp. 1496–500.

[15] Lotfinegad, P., et al., Immunomodulatory Nature and Site Specific Affinity of Mesenchymal Stem Cells: a Hope in Cell Therapy. Adv Pharm Bull, 2014. **4**(1): pp. 5–13.

[30] Qiao, L., et al., Suppression of tumorigenesis by human mesenchymal stem cells in a hepatoma model. Cell Res, 2008. **18**(4): pp. 500–7.

[31] Ohlsson, L. B., et al., Mesenchymal progenitor cell-mediated inhibition of tumor growth in vivo and *in vitro* in gelatin matrix. Exp Mol Pathol, 2003. **75**(3): pp. 248–55.

[32] Zhu, Y., et al., Human mesenchymal stem cells inhibit cancer cell proliferation by secreting DKK-1. Leukemia, 2009. **23**(5): pp. 925–33.

[33] Stupp, R., et al., Radiotherapy plus concomitant and adjuvant temozolomide for glioblastoma. N Engl J Med, 2005. **352**(10): pp. 987–96.

[34] Stoicov, C., et al., Mesenchymal stem cells utilize CXCR4-SDF-1 signaling for acute, but not chronic, trafficking to gastric mucosal inflammation. Dig Dis Sci, 2013. **58**(9): pp. 2466–77.

[35] Yagi, H. and Y. Kitagawa, The role of mesenchymal stem cells in cancer development. Front Genet, 2013. **4**: pp. 261.

[36] Rattigan, Y., et al., Interleukin 6 mediated recruitment of mesenchymal stem cells to the hypoxic tumor milieu. Exp Cell Res, 2010. **316**(20): pp. 3417–24.

[37] Nakamura, K., et al., Antitumor effect of genetically engineered mesenchymal stem cells in a rat glioma model. Gene Ther, 2004. **11** (14): pp. 1155–64.

[38] Nakamizo, A., et al., Human bone marrow-derived mesenchymal stem cells in the treatment of gliomas. Cancer Res, 2005. **65**(8): pp. 3307–18.

[39] Bexell, D., et al., Bone marrow multipotent mesenchymal stroma cells act as pericyte-like migratory vehicles in experimental gliomas. Mol Ther, 2009. **17**(1): pp. 183–90.

[40] Bexell, D., S. Scheding, and J. Bengzon, Toward brain tumor gene therapy using multipotent mesenchymal stromal cell vectors. Mol Ther, 2010. **18**(6): pp. 1067–75.

[41] Munoz, J. L., et al., Delivery of Functional Anti-miR-9 by Mesenchymal Stem Cell-derived Exosomes to Glioblastoma Multiforme Cells Conferred Chemosensitivity. Mol Ther Nucleic Acids, 2013. **2**: p. e126.

[42] Kim, S. M., et al., Effective combination therapy for malignant glioma with TRAIL-secreting mesenchymal stem cells and lipoxygenase inhibitor MK886. Cancer Res, 2012. **72**(18): pp. 4807–17.

[43] Behnan, J., et al., Recruited brain tumor-derived mesenchymal stem cells contribute to brain tumor progression. Stem Cells, 2013.

[44] Akimoto, K., et al., Umbilical cord blood-derived mesenchymal stem cells inhibit, but adipose tissue-derived mesenchymal stem cells promote,

glioblastoma multiforme proliferation. Stem Cells Dev, 2013. **22**(9): pp. 1370–86.

[45] Martinez-Quintanilla, J., et al., Therapeutic efficacy and fate of bimodal engineered stem cells in malignant brain tumors. Stem Cells, 2013. **31**(8): pp. 1706–14.

[46] Friedenstein, A. J., R. K. Chailakhjan, and K. S. Lalykina, The development of fibroblast colonies in monolayer cultures of guinea-pig bone marrow and spleen cells. Cell Tissue Kinet, 1970. **3**(4): pp. 393–403.

[47] Friedenstein, A. J., J. F. Gorskaja, and N. N. Kulagina, Fibroblast precursors in normal and irradiated mouse hematopoietic organs. Exp Hematol, 1976. **4**(5): pp. 267–74.

[48] Angelopoulou, M., et al., Cotransplantation of human mesenchymal stem cells enhances human myelopoiesis and megakaryocytopoiesis in NOD/SCID mice. Exp Hematol, 2003. **31** (5): pp. 413–20.

[49] Chen, S., et al., Intracoronary transplantation of autologous bone marrow mesenchymal stem cells for ischemic cardiomyopathy due to isolated chronic occluded left anterior descending artery. J Invasive Cardiol, 2006. **18**(11): pp. 552–6.

[50] Chen, S. L., et al., Effect on left ventricular function of intracoronary transplantation of autologous bone marrow mesenchymal stem cell in patients with acute myocardial infarction. Am J Cardiol, 2004. **94**(1): pp. 92–5.

[51] Maltman, D. J., S. A. Hardy, and S. A. Przyborski, Role of mesenchymal stem cells in neurogenesis and nervous system repair. Neurochem Int, 2011. **59**(3): pp. 347–56.

[52] Liu, L., et al., From blood to the brain: can systemically transplanted mesenchymal stem cells cross the blood-brain barrier? Stem Cells Int, 2013. 2013: p. 435093.

[53] Uccelli, A., et al., Neuroprotective features of mesenchymal stem cells. Best Pract Res Clin Haematol, 2011. **24**(1): pp. 59–64.

[54] Crigler, L., et al., Human mesenchymal stem cell subpopulations express a variety of neuro-regulatory molecules and promote neuronal cell survival and neuritogenesis. Exp Neurol, 2006. **198**(1): pp. 54–64.

[55] Wilkins, A., et al., Human bone marrow-derived mesenchymal stem cells secrete brain-derived neurotrophic factor which promotes neuronal survival *in vitro*. Stem Cell Res, 2009. **3**(1): pp. 63–70.

[56] Patel, N., et al., Developmental regulation of TAC1 in peptidergic-induced human mesenchymal stem cells: implication for spinal cord injury in zebrafish. Stem Cells Dev, 2012. **21**(2): pp. 308–20.

have offered promising strategies for reconstructing and repairing defective tissues in vivo (1-2), enabling damaged tissue to be replaced with cultured tissues that meet the needs of the individual patients. A number of companies manufacturing cultured tissues have been established. The manufacture of cultured tissues is still burdened by instability owing to the qualitative fluctuation of cell sources as raw materials and the risk of biological contamination.

Innovative techniques of cell and tissue processing have been developed for therapeutic applications. The subculturing for cell expansion is a core process. In manufacturing, strict management against contamination and human error are compelled due to un-sterilable products and the complexity of culture techniques, respectively. In addition, the development of a processing system is considered to lead to safety, security and cost-saving for cell and tissue cultures. However, the criterion of facility design to date has not been clear. This article describes a novel strategy for bioreactor and facility designs.

7.2 Bioreactor Design for Cell Processing

Bioreactors are a core element to produce high-value materials in biological processes using mammary cells which can be employed for many purposes on various scales of operation in pharmaceutical production, cell therapies and tissue engineering (3). These range from simple, small-scale systems for basic research to sophisticated production-scale systems, which are in use for commercial manufacture. The development of industrial-scale bioreactors was initiated in the mid-1950s to meet the demand for mass production of vaccines. The cell culture bioreactors employed stirred tanks containing micro-carriers with adherent cells, which were, in principle, an adaptation of homogeneous culture systems used for microbial culture to meet the requirements of mechanically sensitive animal cells. The fundamental idea was to overcome the major limitations of cell cultivations that caused slow cell growth and low attainable cell densities by providing an environment that allowed the cells to continuously produce the products of interest at high levels.

In recent years, a new trend has emerged, that of tissue engineering. In contrast to traditional approaches of bioreactor design for mass production, the manufacturing features inherent for cell-based health care products leads to the requirement for small-scale design of patient-oriented bioreactors for clinical use. The automation platform becomes the core technology to realize the 3S (safety, security and cost-saving of manufacturing). Especially, the installation of a processing system for cell and tissue cultures leads to: 1) process automation of machinery operations, 2) maintenance of the closed aseptic environment to reduce contamination risks, 3) mimicry of the biological environment with chemical and mechanical stimuli, and 4) information utilization of culture monitoring. These functional progresses provide some solutions to the features inherent to cell and tissue processing, being contributable not only to the process control including the saving of labor and process stability, but also to the quality control including the evaluation of cell and tissue potentials.

In our previous study, the automation for expansion process including the operations of seeding, medium change, passage as well as observation were developed (4, 5), and proposed the intelligent culture system accompanied by automated operations (liquid transfer and cell passage) to perform serial cultures of human skeletal muscle myoblasts, as shown in Figure 7.1(6). An automated culture system that could manage two serial cultures by monitoring the confluence degree was constructed. The automated operation with the intelligent determination of the time for passage was successfully performed without serious loss of growth activity, compared with manual operation using conventional flasks. This intelligent culture system can be applied to cultures of other adherent cells and will lead to the qualitative stability of products in the practical manufacturing of cells available for transplantation.

Recently, this technique applied to the development of chip culture system for maturation of retina pigment epithelial cells derived from human iPS cells (Figure 7.2). The chip bioreactor system for long-term culture of human retina pigment epithelial (RPE) cells consists of incubation unit and medium supplier unit. In the incubation unit, the chip as closed vessel (2.5 mm in diameter, working volume 25 ml) was set to be 37°C and 5% CO_2, where gas permeable resin (PDMS) was used for the vessel wall. Whole bottom surface of chip was observed through the culture to detect the immature

Figure 7.1 Intelligent bioreactor system for passage automation. A; overview of the system, B; culture vessel, and C; monitoring system.

In the pharmaceutical manufacturing of healthcare products, the minimization of space, operator's entrance and cross-contamination raise the development of the isolator as useful alternatives to full-scale clean rooms, being described as: "A device creating a small, enclosed, controlled or clean-classified environment in which a process or activity can be placed with a high degree of assurance that effective segregation will be maintained between the closed environment, its surroundings and any personnel involved with the process or manipulation" (14). According to the ISO guideline for the aseptic processing of healthcare products (Part 6: Isolator systems, ISO 13408-6), the isolator is placed in a clean room in which the environment is controlled to give the same conditions as an ISO Class 8 clean area equivalent to an indirect supporting zone in the aseptic processing of healthcare products. An economic analysis using the parameter of lifecycle cost indicated that the total cost per lot in the infrastructures for aseptic cell processing was based on: (i) the critical processing zone with manual operations, (ii) the isolator with manual operations, and (iii) the isolator with automated operations using the robot arm. Aseptic cell processing based on the isolator system with manual operations could reduce the lifecycle cost by 43%, compared with that based on the critical processing zone (15). The installation of a robotic system to realize automated processing in the isolator was suggested to achieve a 38% reduction in cost in the production scale, although the expenses related to facility costs increased by 2% compared with that based on the critical processing zone. Even though a further estimation will be required for practical management of the aseptic processing of cells and tissues for therapeutic use, these estimations are considered to promote the broad utility of the isolator for the aseptic cell processing for not only healthcare products but also for therapeutic cells and tissues.

The installation of isolator technology applied to the cell and tissue processing for therapeutic application would be a similar layout to that for aseptic processing of healthcare products as mentioned above. The critical issue of the isolator is to equip the pass box with a decontamination apparatus, so that the aseptic environment can be prepared by exposing it to decontamination reagents such as vaporized hydrogen peroxide using the decontamination apparatus, enabling materials such as culture vessels and containers for cells and medium to pass through the border from the ISO Class 8 clean area (equivalent to indirect supporting zone) to the ISO Class 5 clean area (equivalent to critical processing zone) without any additional buffer spaces (equivalent to the ISO Class 7 clean area of direct support zones), enabling the saving of costs for the operation and maintenance as well as space in the manufacturing of cell and tissues. As the siting criteria depend on the

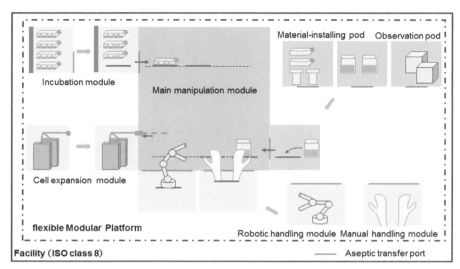

Figure 7.3 Proposal of manufacturing system based on flexible Modular Platform (fMP)

Figure 7.4 Automation system of sheet assembly based on the fmp technology

between modules under aseptic conditions by developing the interface of double door system for modules, suggesting the broad versatility for the production in other types of multilayered cell sheets.

7.5 Acknowledgments

This study was supported by the Japan Society for the Promotion of Science (JSPS) through the "Funding Program for World-Leading Innovative R&D on Science and Technology (FIRST Program)," initiated by the Council for

Science and Technology Policy (CSTP) and by the Strategic Promotion of Innovative Research and Development (S-Innovation) program of the Japan Science and Technology Agency (JST),and by the project of "Development of cell manufacturing and processing system for industrialization of regenertive medicine"(No.P14006)commissioned by the New Energy and Industrial Technology Development Organization(NEDO).

References

[1] Mason, C. and Hoare, M.: Regenerative medicine bioprocessing: building a conceptual framework based on early studies. Tissue Eng., 13, 301–311 (2007).

[2] Hesse, F. and Wagner, R.: Developments and improvements in the manufacturing of human therapeutics with mammalian cell cultures. Trends Biotechnol., 18, 173–180 (2000).

[3] Taya, M. and Kino-oka, M.: Bioreactors for animal cell cultures, Comprehensive Biotechnology, 2nd edition (eds. by M. Butler, C. Webb, A. Moreira, B. Grodzinski, Z. F. Cui, S. Agathos, M. Moo-Young), Vol.2, pp. 373–382, Elsevier (2011)

[4] Kino-oka, M. and Prenosil, J. E.: Development of an on-line monitoring system of human keratinocyte growth by image analysis and its application to bioreactor culture, Biotechnol. Bioeng., 67, 234–239 (2000).

[5] Kino-oka, M., Ogawa, N, Umegaki, R., and Taya, M.: Bioreactor design for successive culture of anchorage-dependent cells operated in an automated manner. Tissue Eng., 11, 535–545 (2005).

[6] Kino-oka, M., Chowdhury, S. R., Muneyuki, Y., Manabe, M., Saito, A., Sawa, Y., and Taya, M.: Automating the expansion process of human skeletal muscle myoblasts with suppression of myotube formation, Tissue Eng., 15, 717–728 (2009).

[7] Thomas, R. J. Hourd, P. C., and Williams, D. J.: Application of process quality engineering techniques to improve the understanding of the *in vitro* processing of stem cells for therapeutic use. J. Biotechnol., 136, 148–155 (2008).

[8] Portner, R., Nagel-Heyer, S., Goepfert, C., Adamiez, P., and Meenen, N. M.: Bioreactor design for tissue engineering. J. Biosci. Bioeng., 100, 235–245 (2005).

8.2 Evidence for the Existance of Melanoma Stem Cells with Self-Renewing and Tumorigenic Properties

During the past decades, indirect evidence has supported the presence of melanoma stem cells. First, melanomas show phenotypic heterogeneity both *in vitro* and *in vivo*, suggesting an origin from a cell with multilineage differentiation abilities. Melanomas retain their morphologic and biological plasticity, despite repeated cloning. Second, melanoma cells often express developmental genes such as Sox10, Pax3, Mitf and Nodal. Melanomas also express the intermediate filament Nestin, which is associated with multiple stem cell populations. Third, melanoma cells can differentiate into a wide range of cell lineages, including neural, mesenchymal and endothelial cells.

Next to these indirect findings, recent studies have provided direct evidence for the existence of melanoma stem cells [reviewed in 2]. Applying growth conditions suitable to human embryonic stem cells, Fang and colleagues [3] found a subpopulation of melanoma cells propagating as non-adherent spheres in approximately 20% of metastatic melanomas, whereas in standard media only adherent monolayer cultures developed. Sphere formation *in vitro* has been proposed by different groups as a common growth feature of stem cells, including neural crest–derived stem cells. The authors showed that melanoma spheres can differentiate under appropriate culture conditions into multiple lineages, such as melanocytes, adipocytes, osteocytes, and chondrocytes, recapitulating the plasticity of neural crest stem cells [3]. Multipotent melanoma spheroid cells persisted over several months after serial cloning *in vitro* and transplantation *in vivo*, indicating a stable capacity to self-renew. Interestingly, sphere cells were more tumorigenic than their adherent counterparts when grafted into mice. Finally, the authors found that the stemness criteria were significantly enriched in a small CD20-positive subpopulation, indicating that CD20 might be a suitable surface marker for the identification of melanoma stem cells [3].

Additional support for the existence of melanoma stem cells came from the finding that the surface marker CD133, a stem cell marker previously applied to neural stem cells, could be employed to isolate a subset of stem-like melanoma cells from patient biopsies [4]. Using fluorescence-activated cell sorting (FACS) from freshly isolated melanoma cells, the authors demonstrated that the CD133-positive subpopulation represents less than 1% of the total tumor mass of melanoma, a finding consistent with designated stem cell subpopulations from other tissues. Like the CD20-positive population defined by Fang and colleagues [3], CD133-positive melanoma cells revealed an

Additional evidence that melanoma follows a cancer stem cell model came from recent studies. Melanoma cells from primary and metastatic tumors that expressed CD271 (nerve growth factor receptor), a surface marker of neural crest stem cells, displayed a marked capacity for self-renewal and differentiation plasticity when compared with CD271- melanoma cells. Furthermore, when tumorigenesis was examined CD271+ melanoma cells, but not CD271-, formed tumors and metastases [9, 10]. Interestingly, CD271 and ABCB5 were recently found to be co-expressed in clinical human melanoma samples.

In another study, a sub-population of melanoma cells identified by higher aldehyde dehydrogenase (ALDH) activity displayed enhanced tumorigenicity and capacity to self-renewal when compared to ALDH-negative cells [11]. Melanoma cells that express the receptor activator of NK-kB (RANK) demonstrated increased tumorigenicity compared with RANK-negative melanoma cells. Moreover, RANK was coexpressed with ABCB5 and CD133 in melanoma cells, and preferentially expressed by peripheral circulating melanoma cells [12]. Recently, a temporary distinct subpopulation of slow-cycling cells characterized by the expression of the H3K4 demethylase JARID1B has been shown to be required for continuous melanoma growth [13].

8.3 The Hedgehog Signaling Pathway

A handful of morphogenetic signaling pathways regulating developmental processes, organ homeostasis and self-renewal in normal stem cells, plays also a critical role in tumorigenesis. Among them, Hedgehog (Hh) is crucial for determining proper embryonic patterning and controlling growth and cell fate during animal development. Similarly, in the adult, is involved in tissue maintenance and repair, regulating stem cell behaviour.

Activation of the Hh signaling is initiated by the binding of Hh ligands (Sonic, Indian and Desert) to the trans-membrane protein Patched (Ptch), which, in absence of the ligands, represses the pathway by preventing the activation of the essential trans-membrane protein Smoothened (Smo). Binding of Hh to Ptch allows activation of Smo, leading to the formation of activating forms of the Gli zinc finger transcription factors Gli1, Gli2 and Gli3 [14, 15] (Figure 8.1). Direct transcriptional activation of Gli1 by Gli2/3 enhances the level of Gli activators and high level expression of Gli1 is considered a reliable indicator of Hh pathway activity. Gli1 and Gli2 act as main mediators of Hh signaling in cancer by controlling the expression of target genes involved in proliferation, metastasis, survival and

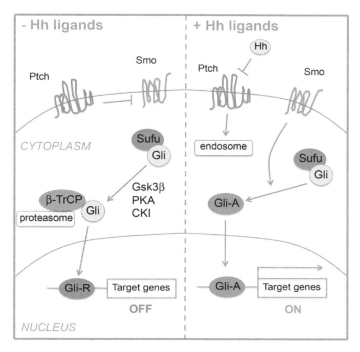

Figure 8.1 Schematic diagram of the Hh signaling pathway. In the absence of the ligand (left), the Ptch receptor suppresses the function of Smo. Full-length Gli proteins (Gli, yellow) are converted to a C-terminally truncated repressor form (Gli-R, red). Formation of the Gli-R is promoted by sequential phosphorylation of full-length Gli by GSK3 β, PKA and CKI, which creates binding sites for the adapter protein β-TrCP, becoming subject of ubiquitination. The Gli-R mediates transcriptional repression of target genes. In presence of the ligands (right), binding inhibits Ptch's function, which results in activation of Smo. Active Smo promotes the activation of full-length Gli proteins (Gli, green), which enters the nucleus and promotes transcription of target genes.

stemness [14, 15]. The activity of the three Gli proteins is tightly controlled. First, nuclear-cytoplasmic shuttling is tightly regulated by protein kinase A (PKA) and by Suppressor of Fused (Sufu), which not only prevents their nuclear translocation, but also inhibits Gli1-mediated transcriptional activity. Second, ubiquitination and protein degradation of Gli proteins are regulated by several distinct mechanisms, including β-TrCP-, cul3/BTB-, Numb/Itch- and acetylation-mediated ubiquitination. Third, Gli3 and, to a lesser extent, Gli2, can be processed into transcriptional repressors [reviewed in 16] (Figure 8.1).

study highlights the role of the HH signaling pathway in driving self-renewal and tumorigenicity of melanoma stem cells and points to SMO and GLI1 as novel and effective therapeutic targets for the treatment of human melanoma.

Although most reported findings seem to be highly promising, many unanswered questions still exist. We do not yet know how many subpopulations of melanoma cells with stem cell properties exist. First, is there a definite number of clearly distinguishable subpopulations, or is there a continuous spectrum of cells, that is passing through a state of trans-amplifying cells that gradually lose their stemness, to differentiated tumor cells? Second, although many cancers contain cells that display stem cell–like features, the identity of the normal cell that acquires the first genetic hit leading to the tumor-initiating cell remains elusive in melanoma. Normal cells that already have stem cell properties represent likely targets, but other mechanisms are conceivable. Third, almost nothing is known so far about the niche of melanoma stem cells. What is the impact of the niche on melanoma development, maintenance and metastasis? As a long-term perspective, melanoma stem cell research will certainly influence and improve the diagnosis, prognosis, and therapy of melanoma. Traditional treatments might be recalibrated and novel therapies need to be developed focusing on the ability to target the melanoma stem cell population and its specific signaling pathways.

8.6 Acknowledgement

Work in the authors laboratory has been supported by Associazione Italiana per la Ricerca sul Cancro (AIRC, 9566), Regional Health Research Program 2009 and Fondazione Cassa di Risparmio di Firenze.

References

[1] Ko J. M., Fisher D. E. A new era: melanoma genetics and therapeutics. *J Pathol.* 223:241–250, 2011.

[2] Girouard S. D., Murphy G. F. Melanoma stem cells: not rare, but well done. *Lab Invest.* 91:647–664, 2011.

[3] Fang D., Nguyen T. K., Leishear K. et al. A tumorigenic subpopulation with stem cell properties in melanomas. *Cancer Res.* 65:9328–9337, 2005.

[4] Monzani E., Facchetti F., Galmozzi E. et al. Melanoma contains CD133 and ABCG2 positive cells with enhanced tumorigenic potential. *Eur J. Cancer* 43:935–946, 2007.

[19] Lai K., Kaspar B. K., Gage F. H. et al. Sonic hedgehog regulates adult neural progenitor proliferation *in vitro* and *in vivo*. *Nat Neurosci*. 6: 21–27, 2003.

[20] Hutchin M. E., Kariapper M. S., Grachtchouk M. et al. Sustained Hedgehog signaling is required for basal cell carcinoma proliferation and survival: conditional skin tumorigenesis recapitulates the hair growth cycle. *Genes Dev* 19:214–233, 2005.

[21] Alexaki V. I., Javelaud D., Van Kempen L. C. et al. GLI2-mediated melanoma invasion and metastasis. *J Natl Cancer Inst* 102:1148–1159, 2010.

[22] Clement V., Sanchez P., de Tribolet N. et al. HEDGEHOG-GLI1 signaling regulates human glioma growth, cancer stem cell self-renewal and tumorigenicity. *Curr Biol* 17:165–172, 2007.

[23] Varnat F., Duquet A., Malerba M. et al. Human colon cancer epithelial cells harbour active HEDGEHOG-GLI signaling that is essential for tumor growth, recurrence, metastasis and stem cell survival and expansion. *EMBO Mol Med* 1:338–351, 2009.

[24] Song Z., Yue W., Wei B. et al. Sonic hedgehog pathway is essential for maintenance of cancer stem-like cells in human gastric cancer. *PLoS One* 6:e17687, 2011.

[25] Peacock C. D., Wang Q., Gesell G. S. et al. Hedgehog signaling maintains a tumor stem cell compartment in multiple myeloma. *Proc Natl Acad Sci USA* 104:4048–4053, 2007.

[26] Dierks C., Beigi R., Guo G. R. et al. Expansion of Bcr-Abl-positive leukemic stem cells is dependent on Hedgehog pathway activation. *Cancer Cell* 4:238–249, 2008.

[27] Santini R., Vinci M. C., Pandolfi S. et al. HEDGEHOG-GLI Signaling Drives Self-Renewal and Tumorigenicity of Human Melanoma-Initiating Cells. *Stem Cells* 30:1808–18, 2012.

9

A Quest for Refocussing Stem Cell Induction Strategies: How to Deal with Ethical Objections and Patenting Problems

Hans-Werner Denker

Lehrstuhl für Anatomie und Entwicklungsbiologie,
Universität Duisburg-Essen, Germany*
Corresponding author: Hans-Werner Denker
<hans-werner.denker@uni-due.de>

Abstract

A recent ruling of the European Court of Justice (EU-CJ) in Luxembourg (18 October 2011) on ES cell patenting has renewed the interest in addressing so-far unsolved ethical problems of stem cell research. In this contribution I will outline ethical and patenting problems that arise when working with pluripotent stem cells, specifically in the modern field of induced pluripotent stem cell (iPSC) technology. The focus will be on stem cell potentiality, and I will argue that potentiality rather than the act of sacrificing embryos will have to be a central point of concern in stem cell ethics and patenting in the future. Possible solutions will be discussed.

When somatic cells are reprogrammed to gain "full" pluripotency, they acquire (so to say as a by-product) the capability to form viable embryos if tetraploid complementation (TC) is performed (termed "gold standard" by some authors). I argue that any human cells possessing this capability cannot be patented. In analogy to the arguments used by the EU-CJ, this must apply not only to patenting cell lines themselves but also to patenting technologies using these cells. The patenting problem is more than an obstacle for researchers and

---

This contribution reflects a lecture given at the Symposium '3rd Disputationes on Native and Induced Pluripotent Stem Cell Standardization', March 19-21, 2012, Florence, Italy.
*web: http://www.uni-due.de/denker/

Mayuri Prasad and Paolo Di Nardo (Eds.), Innovative Strategies in Tissue Engineering, 117–134.

9.2 Potential for Autonomous Pattern Formation: Embryoid Bodies

One of the remarkable biological properties of pluripotent cells is their ability to form, in suspension cultures, **"embryoid bodies" (EBs)**. What is most widely known about EBs is that formation of these embryo-like structures promotes the formation of germ layers. Much less intensely studied is the degree of order that the germ layers and their derivatives can attain in EBs, and how close their organisation can really come to the basic body plan of viable embryos. Surprisingly, this aspect appears to receive increased attention only recently.

A particularly remarkable observation on early embryonic pattern formation in EBs was published already in one of the pioneering papers on ESCs [7]. In this case the spontaneous formation of astonishingly embryo-like structures was observed in dense cultures of common marmoset ESCs (Callithrix jacchus, a South American primate): These structures were described to consist of a flat embryonic disc as typical for primates, with primitive ectoderm, primitive endoderm, an amnion with amniotic cavity, a yolk sac. Most remarkably, those authors also depicted and described, within this embryonic disc, an area of ordered ingression of cells, which they addressed as a primitive streak (PS). The PS is of utmost importance in vertebrate embryology, because it is the site where not only the formation of the definitive germ layers (in particular mesoderm and definitive endoderm) takes place but is also instrumental in individuation: The anterior part of the PS is the equivalent of Spemann's organizer which plays a central role in laying down the basic body plan, i.e. the ordered arrangement of germ layers and their derivatives according to the main body axes (dorso-ventral, anterior-posterior = cranio-caudal). The development of single or double organizers is decisive for the formation of a singlet vs. monozygotic twins (discussed in [8]).

While the structures formed spontaneously in the marmoset ESC cultures according to Thomson et al. [7] were remarkably embryo-like, and even appeared to show signs of incipient individuation (discussed in [8]) a comparable degree of order has never since been described to occur in ESC cultures in any other species, including the rhesus monkey and the mouse. Locally restricted gastrulation-like events have, however, been observed [9–12]. The degree of order attained appears to depend very much on physical conditions of culturing [9, 12], as was to be expected on the basis of what we know from developmental biology [8]. It is strongly influenced by other (non-stem) cells or matrix simultaneously present in the cultures. On the other

TC was developed as a variant of chimera formation in the mouse. It provides a method to produce viable embryos and even offspring derived entirely from PSCs that had been propagated before *in vitro* [13–15]. The method relies on the combination of the pluripotent cells with tetraploidized blastomeres or, alternatively, on the injection into tetraploidized blastocysts. Remarkably, this method of cloning viable individuals is successful not only with ESCs but also with iPSCs. In the latter case the term **"all-iPSC mice"** has become popular for the products of this type of cloning [16–20].

Cloning by TC is widely used in experimental research as some type of quality control for PSCs (ESCs and iPSCs) in the mouse. Testing cells for TC capability is being addressed by some authors as the "gold standard" of pluripotency and its use is being advocated in a way that might suggest using it also with human cells [16, 18].

Why would anyone possibly be interested in applying TC with **human** cells? It may appear improbable that TC will be used in the near future for **reproductive cloning** in the human since so far there is widespread consensus to consider this unethical. However, there appears to be reason to question whether this consensus can be expected to hold for long and worldwide. In the Western world, it has already been proposed to consider using TC technology in order to increase success rates in *in-vitro* fertilization/embryo transfer (IVF-ET) clinics [21]. The idea is to derive ESCs from human IVF embryos, expand them *in vitro*, and (re-)construct from them a (larger) number of embryos by TC which can then be used for embryo transfer (while aliquots of the ESCs as well as some of the numerous identical embryos produced could of course be stored in liquid nitrogen to be available for later use in repeated attempts) [21]. Since this means cloning, and since reproductive cloning in the human is considered illegal at least in a number of countries, one may be skeptical whether this technique will ever be applied in Western world countries. However, it cannot be excluded that legislation may develop in a different direction in other cultural environments. As an example, Buddhist authorities have expressed that they would consider embryo destruction in the course of "therapeutic cloning" (for the production of ESCs) unethical, but not so the (re-)construction of embryos in the course of reproductive cloning (for literature see [4]).

While any application of TC for reproductive cloning in the human may appear improbable at the moment its use for **research** and **quality testing** purposes has indeed been proposed frequently, in the mouse, and with respect to human cells at least indirectly. This notion is found in the literature particularly often since the advent of iPSC research. First of all TC is recommended as the most rigorous pluripotency test ("gold standard") for iPS cells in the mouse (*"We therefore consider the tetraploid complementation as the state-of-the-art technique to assess the pluripotency of a given cell line"* [22]; *"This study underscores the intrinsic qualitative differences between iPS cells generated by different methods and highlights the need to rigorously characterize iPS cells beyond in vitro studies."* [23]). Likewise in the first reports on the generation of viable mice from iPSCs it had already been suggested indirectly to apply TC technology for iPSC quality testing also in the human, for the reason that this is considered the most rigorous pluripotency test [16, 18]. Remarkably, it was felt necessary, therefore, to publish a comment on these papers (in the same journal) clarifying that for ethical reasons it cannot be defended to follow this (implicit) recommendation, i.e. to use the technique for iPSC quality testing in the human [5].

Temptations may indeed be high to consider applying TC technology with human pluripotent cells, in spite of these warnings. Why? Recent literature is full of data asking for stringent quality testing in iPSC research. Individual iPSC lines are observed, in the mouse as well as in the human, to vary with respect to differentiation capacities, gene expression patterns and epigenetic marks/memory [24–28]. Stadtfeldt et al. [20] provided a typical and interesting example. They observed that transcripts encoded within the imprinted Dlk1-Dio3 gene cluster were aberrantly silenced in most of the iPSC clones, and that these clones failed to support the development of entirely iPSC-derived animals ("all-iPSC mice") when TC was performed thus revealing a lack of "complete pluripotency". This failure could, however, be corrected by a treatment with a histone deacetylation inhibitor which reactivated the locus. It is clear that investigators wish to have a test available to monitor the success of this type of cell quality improvement. In addition to epigenetic peculiarities of PSC lines (as compared to early embryonic cells) even chromosomal aberrations and gene deletions have been observed in some cases [29, 30].

Such observations obviously could be seen as a strong argument for using the most stringent pluripotency test (TC) also with human iPSCs in order to select "optimal" cell lines and/or stem cell derivation protocols. This logic may be particularly obvious if cells are to be used for **therapeutic purposes** (cell and tissue replacement) in the human, and likewise whenever iPSCs are used for **disease modelling**. In cell and tissue replacement, the concern is that transplanted cells should be genetically and epigenetically as "normal" as possible in order to minimize risks e.g. with respect to tumor formation. In disease modelling with genetic/epigenetic focus, experiments are usually done in the first place in the mouse model, sometimes including TC technology

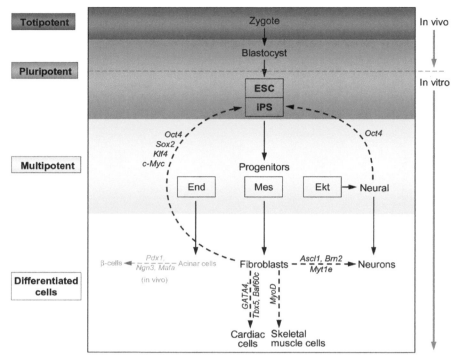

Figure 9.1   Strategies for stem and progenitor cell derivation and cell reprogramming. The *"traditional" strategy* includes a pluripotent state of cells (ESC, iPSC; above); from these pluripotent stem cells multipotent lineage-specific progenitors and finally the various differentiated cell types are derived. In case of iPSC, the cells of origin are differentiated cells (e.g. fibroblasts) which are reprogrammed by activation of the four pluripotency-associated genes Oct4, Sox2, Klf4 and c-Myc (Yamanaka factors; left part of diagram). The *alternative strategy* (lower part of the diagram) avoids the induction of the ethically problematic pluripotent state (*direct reprogramming, bypassing pluripotency*): In this strategy, transcription factor-induced lineage reprogramming results either in cells remaining within the same cell lineage (i.e. mesoderm), or may produce functional cells of other lineages (converting mesodermal fibroblasts into ectodermal neurons). This transcription factor-induced lineage reprogramming not only avoids the ethical problem posed by a self-organizing and cloning capability gained (e.g. TC capability) but possibly also reduces tumor-formation risks (from [48] with permission).

(ESCs and iPSCs) offer the advantage of growing well and of being able, in addition, to differentiate as desired for regenerative medicine. In the "classical" approach, iPSCs are created by transduction and overexpression or at least temporal activation of the "Yamanaka factor" genes Oct4, Sox2,

reprogramming/bypassing pluripotency mentioned in the previous paragraph suggest that we have now arrived at a point at which such avenues become indeed a most attractive option. One may ask why such a redirection of focus (direct reprogramming) has not been searched for more actively already during previous years, although the ethical arguments why such efforts should appear necessary had already been published since years.

The new strategies of inducing direct conversion of somatic cells to a stem/progenitor state, bypassing pluripotency, appear highly promising and recommendable. They are obviously preferable for ethical reasons because they avoid the problem created by inducing an embryo formation/cloning potential which fully pluripotent cells have. An additional advantage of these new strategies may be to reduce the risk of tumor formation after transplanting such cells because the tumor formation potential may be connected with the embryonic pattern formation/self-organization potential. A word of caution appears to be in place, nevertheless: In order for any such alternative strategy to be ethically acceptable, it must be made sure that it does not involve a transitory state of pluripotency that could remain undetected. Many of the induction protocols require very long culturing time periods, and we are far from understanding exactly what cascade of events takes place during this time period. Some of the protocols include while some omit Oct4, some use combinations of certain (but not all) of the Yamanaka factors while others do not. Which of the possible protocols will be safest in order to exclude TC capability as well as tumor formation risks? It will have to be discussed which genes should be seen here as crucial (e.g. genes involved in early embryonic pattern formation / self-organization processes) [4, 47]. This will be an important topic for future research. Strategies for testing this will need to be developed and discussed: For ethical reasons it cannot be defended to test human cells by cloning via TC [5]. It will thus be necessary to define appropriate combinations of *in vitro* gene expression profiling that may be useful instead in a first approximation, combined with *in vitro* culturing conditions that avoid the initiation of individuation processes. In any case it will be necessary to improve the catalogue of informations routinely given to cell donors: This information needs to include all aspects of the potentiality that the donated cells will or may acquire as a result of reprogramming, including TC capability, because this touches upon personal interests donors have (genetic identity and uniqueness). Implications of cell banking need to be included keeping in mind that ethical and legal standards may change and already differ in the various cultural environments. These aspects are particularly relevant when long term storage and widespread use of the cells are envisaged.

Obviously this complex field of ethical problems can be avoided by circumventing any gain of pluripotency at all. A general recommendation for strategies of stem cell derivation would thus be to deviate from the widespread practice of activating the pluripotency program (creating iPSCs) and rather to rely on the alternative strategies bypassing pluripotency, i.e. to convert cells directly to multipotent stem/progenitor cells as in the examples given above.

9.7 Acknowledgments

I like to thank Anna M. Wobus and John Wiley & Sons for granting me permission to re-publish Figure 9.1 (Figure 2 in [48]).

References

[1] EU-CJ, Judgment of the Court (Grand Chamber) of 18 October 2011 (reference for a preliminary ruling from the Bundesgerichtshof - Germany) - Oliver Brüstle v Greenpeace e.V. (Case C-34/10) 1. EUGH Curia Europa(http://curia.europa.eu/juris/document/document.jsf?docid= 115334&mode=lst&pageIndex=1&dir=&occ=first&part=1&text=& doclang=EN&cid=114275), 2011.

[2] Callaway E. European court bans patents based on embryonic stem cells. Final decision could stifle investments in developing therapies. Nature News (Published online 18 October 2011, doi:10.1038/news.2011.597), 2011

[3] González F., Boué S., Belmonte J. C. Methods for making induced pluripotent stem cells: reprogramming a la carte. Nat. Rev. Genet., 12: 231–242, 2011.

[4] Denker H.-W. Induced pluripotent stem cells: how to deal with the developmental potential. Reprod. Biomed. Online, 19 Suppl 1: 34–37, 2009.

[5] Denker H.-W. Ethical concerns over use of new cloning technique in humans. Nature, 461 (7262): 341, 2009.

[6] Lo B., Parham L., Alvarez-Buylla A., Cedars M., Conklin B., Fisher S., Gates E., Giudice L., Halme D. G., Hershon W., Kriegstein A., Kwok P. Y., Wagner R. Cloning mice and men: prohibiting the use of iPS cells for human reproductive cloning. Cell Stem Cell, 6 (1): 16–20, 2010.

[28] Nazor K. L., Altun G., Lynch C., Tran H., Harness J. V., Slavin I., Garitaonandia I., Muller F. J., Wang Y. C., Boscolo F. S., Fakunle E., Dumevska B., Lee S., Park H. S., Olee T., D'Lima D. D., Semechkin R., Parast M. M., Galat V., Laslett A. L., Schmidt U., Keirstead H. S., Loring J. F., Laurent L. C. Recurrent variations in DNA methylation in human pluripotent stem cells and their differentiated derivatives. Cell Stem Cell, 10 (5): 620–634, 2012.

[29] Mayshar Y., Ben-David U., Lavon N., Biancotti J. C., Yakir B., Clark A. T., Plath K., Lowry W. E., Benvenisty N. Identification and classification of chromosomal aberrations in human induced pluripotent stem cells. Cell Stem Cell, 7 (4): 521–531, 2010.

[30] Laurent L. C., Ulitsky I., Slavin I., Tran H., Schork A., Morey R., Lynch C., Harness J. V., Lee S., Barrero M. J., Ku S., Martynova M., Semechkin R., Galat V., Gottesfeld J., Izpisua Belmonte J. C., Murry C., Keirstead H. S., Park H. S., Schmidt U., Laslett A. L., Muller F. J., Nievergelt C. M., Shamir R., Loring J. F. Dynamic Changes in the Copy Number of Pluripotency and Cell Proliferation Genes in Human ESCs and iPSCs during Reprogramming and Time in Culture. Cell Stem Cell, 8 (1): 106–118, 2011.

[31] Denker U., Denker H.-W. Embryonale Stammzellforschung: Aufklärung notwendig. Problematik der informierten Zustimmung der Spender. Deutsches Ärzteblatt, 102 (13): A892–A893, 2005.

[32] Vrtovec K. T., Vrtovec B. Is Totipotency of a Human Cell a Sufficient Reason to Exclude its Patentability under the European Law? Stem Cells, 25 (12): 3026–3028, 2007.

[33] Denker H.-W. Totipotency/pluripotency and patentability. Stem Cells, 26 (6): 1656–1657, 2008.

[34] Ieda M., Fu J. D., Delgado-Olguin P., Vedantham V., Hayashi Y., Bruneau B. G., Srivastava D. Direct Reprogramming of Fibroblasts into Functional Cardiomyocytes by Defined Factors. Cell, 142 (3): 375–386, 2010.

[35] Szabo E., Rampalli S., Risueno R. M., Schnerch A., Mitchell R., Fiebig-Comyn A., Levadoux-Martin M., Bhatia M. Direct conversion of human fibroblasts to multilineage blood progenitors. Nature, 468 (7323): 521–526, 2010.

[36] Vierbuchen T., Ostermeier A., Pang Z. P., Kokubu Y., Sudhof T. C., Wernig M. Direct conversion of fibroblasts to functional neurons by defined factors. Nature, 463 (7284): 1035–1041, 2010.

[37] Caiazzo M., Dell'anno M. T., Dvoretskova E., Lazarevic D., Taverna S., Leo D., Sotnikova T. D., Menegon A., Roncaglia P., Colciago G.,

[47] Denker, H.-W. Human embryonic stem cells: the real challenge for research as well as for bioethics is still ahead of us. Cells Tissues Organs, 187 (4): 250–256, 2008.

[48] Wobus, A. M. The Janus face of pluripotent stem cells–connection between pluripotency and tumourigenicity. Bioessays, 32 (11): 993–1002, 2010.

10

Constitutive Equations in Finite Viscoplasticity of Nanocomposite Hydrogels

A.D. Drozdov[1,2,] and J. deClaville Christiansen[2]

[1]Department of Plastics Technology, Danish Technological Institute, Taastrup, Denmark
[2]Department of Mechanical and Manufacturing Engineering, Aalborg University, Aalborg, Denmark
Corresponding author: A.D. Drozdov <add@teknologisk.dk>

10.1 Introduction

This paper deals with constitutive modeling of the viscoplastic response of nanocomposite hydrogels under an arbitrary deformation with finite strains.

Hydrogels are three-dimensional networks of polymer chains connected by physical and chemical cross-links. When a hydrogel is brought in contact with water, it swells retaining its structural integrity and ability to withstand large (up to 3000%) deformations. A shortcoming of conventional (chemically cross-linked) gels that restrains their applicability is that these materials become relatively weak and not sufficiently tough in the swollen state. To enhance mechanical properties of hydrogels without sacrifice of their swellability and extensibility, concentration of reversible physical crosslinks is to be increased [1] either by changes of molecular architecture (double-network hydrogels, gels with hydrophilic and hydrophobic chains [2]) or by reinforcement with nanoparticles that serve as effective multi-functional cross-linkers [3, 4].

Mechanical properties of hydrogels have been a focus of attention in the past decade as these materials demonstrate potential for a wide range of applications including biomedical devices, drug delivery carriers, superabsorbent materials, filters and membranes for selective diffusion, sensors for on-line process monitoring, smart optical systems, and soft actuators [5–8].

Mayuri Prasad and Paolo Di Nardo (Eds.), Innovative Strategies in Tissue Engineering, 135–172.

where the dot stands for inner product, and $\top$ denotes transpose. The principal invariants of the Cauchy–Green tensors are denoted as J_1, J_2, J_3.

The reference configuration of the equivalent network (the configuration in which stresses in chain vanish) differs from the initial configuration. Denote by $\mathbf{F}_*$ and $\mathbf{F}_e$ deformation gradients for transition from the initial configuration into the reference configuration and from the reference configuration into the actual configuration, respectively (the subscript index "e" designates elastic deformation). These tensors are connected with deformation gradient for macro-deformation $\mathbf{F}$ by the multiplicative decomposition formula

$$\mathbf{F} = \mathbf{F}_e \cdot \mathbf{F}_*. \tag{10.2}$$

Transition from the initial configuration into the reference configuration reflects two processes: (i) changes in specific volume (swelling and shrinkage) induced by solvent transport, and (ii) viscoplastic deformation (sliding of junctions between strands in the equivalent polymer network and slippage of nanoparticles with respect to their positions in the initial configuration).

Local transformation of the initial configuration into the reference configuration due to solvent diffusion is described by the deformation gradient $\mathbf{f}$. For an isotropic equivalent medium,

$$\mathbf{f} = f^{\frac{1}{3}}\mathbf{I}, \tag{10.3}$$

where f stands for the coefficient of inflation induced by solvent uptake, and $\mathbf{I}$ is the unit tensor. Local transformation reflecting irreversible sliding is described by the deformation gradient $\mathbf{F}_p$ where the subscript index "p" refers to plastic flow. The plastic deformation is presumed to be volume-preserving,

$$\det \mathbf{F}_p = 1. \tag{10.4}$$

It follows from the multiplicative decomposition formula that [21]

$$\mathbf{F}_* = \mathbf{F}_p \cdot \mathbf{f}. \tag{10.5}$$

Equations (10.2), (10.3), (10.5) imply that

$$\mathbf{F} = f^{\frac{1}{3}}\mathbf{F}_e \cdot \mathbf{F}_p. \tag{10.6}$$

The Cauchy–Green tensors for elastic deformation read

$$\mathbf{B}_e = \mathbf{F}_e \cdot \mathbf{F}_e^\top, \qquad \mathbf{C}_e = \mathbf{F}_e^\top \cdot \mathbf{F}_e. \tag{10.7}$$

It follows from Equations (10.7) and (10.14) that

$$\mathbf{B}_{\mathrm{e}} : \mathbf{D}_{\mathrm{p}} = \mathbf{C}_{\mathrm{e}} : \mathbf{d}_{\mathrm{p}}, \quad \mathbf{I} : \mathbf{D}_{\mathrm{p}} = 0, \quad \mathbf{B}_{\mathrm{e}}^{-1} : \mathbf{D}_{\mathrm{p}} = \mathbf{C}_{\mathrm{e}}^{-1} : \mathbf{d}_{\mathrm{p}}.$$

Insertion of these expressions into Equation (10.16) results in

$$\dot{J}_{\mathrm{e}1} = 2\mathbf{B}_{\mathrm{e}} : \mathbf{D} - 2\mathbf{C}_{\mathrm{e}} : \mathbf{d}_{\mathrm{p}} - \frac{2\dot{f}}{3f} J_{\mathrm{e}1},$$

$$\dot{J}_{\mathrm{e}2} = -2(\mathbf{B}_{\mathrm{e}}^{-1} : \mathbf{D} - \mathbf{C}_{\mathrm{e}}^{-1} : \mathbf{d}_{\mathrm{p}}) J_{\mathrm{e}3} + 2\left(\mathbf{I} : \mathbf{D} - \frac{2\dot{f}}{3f}\right) J_{\mathrm{e}2},$$

$$\dot{J}_{\mathrm{e}3} = 2\left(\mathbf{I} : \mathbf{D} - \frac{\dot{f}}{f}\right) J_{\mathrm{e}3}. \tag{10.17}$$

Denote by

$$\mathbf{B}_{\mathrm{p}} = \mathbf{F}_{\mathrm{p}} \cdot \mathbf{F}_{\mathrm{p}}^{\top}, \qquad \mathbf{C}_{\mathrm{p}} = \mathbf{F}_{\mathrm{p}}^{\top} \cdot \mathbf{F}_{\mathrm{p}} \tag{10.18}$$

the Cauchy–Green tensors for plastic deformation, and by $J_{\mathrm{p}1}$, $J_{\mathrm{p}2}$, and $J_{\mathrm{p}3} = 1$ their principal invariants. Keeping in mind that $\mathbf{d}_{\mathrm{p}}$ is a traceless tensors, we write, by analogy with Equation (10.17),

$$\dot{J}_{\mathrm{p}1} = 2\mathbf{B}_{\mathrm{p}} : \mathbf{d}_{\mathrm{p}}, \quad \dot{J}_{\mathrm{p}2} = -2\mathbf{B}_{\mathrm{p}}^{-1} : \mathbf{d}_{\mathrm{p}}. \tag{10.19}$$

10.2.2 Free Energy Density of a Hydrogel

Denote by Ψ the specific free energy of a nanocomposite hydrogel (per unit volume in the initial configuration). For a hydrogel with an isotropic polymer network, Ψ is treated as a function of seven arguments

$$\Psi = \Psi(J_{\mathrm{e}1}, J_{\mathrm{e}2}, J_{\mathrm{e}3}, J_{\mathrm{p}1}, J_{\mathrm{p}2}, n, t), \tag{10.20}$$

where n stands for numbers of water molecules per unit volume of a hydrogel in its initial state. An explicit dependence of Ψ on time t is introduced to account for evolution of the equivalent polymer network driven by swelling–shrinkage of a nanocomposite hydrogel. The following equation is adopted for the specific free energy

$$\Psi = \mu_0 n + \Psi_{\mathrm{solid}} + \Psi_{\mathrm{mix}}, \tag{10.21}$$

where μ_0 is chemical potential per solvent molecule in the bath (which, in general, differs from chemical potential μ per solvent molecule in a gel),

Ψ_{solid} denotes strain energy density of the solid phase, Ψ_{mix} stands for the energy of mixing of solvent molecules with chains and nanoparticles in the equivalent network.

The strain energy density of an isotropic equivalent medium reads

$$\Psi_{\text{solid}} = \Psi_{\text{solid}}(J_{e1}, J_{e2}, J_{e3}, J_{p1}, J_{p2}, t). \tag{10.22}$$

Within the Flory–Huggins theory of mixing, the specific energy of mixing is given by

$$\phi_s \Psi_{\text{mix}} = \frac{k_B T}{v}(\phi_f \ln \phi_f + \chi \phi_s \phi_f), \tag{10.23}$$

where k_B is Boltzmann's constant, T stands for absolute temperature, v is the characteristic volume of a solvent molecule, χ denotes the Flory–Huggins interaction parameter, and

$$\phi_f = \frac{nv}{1+nv}, \qquad \phi_s = \frac{1}{1+nv} \tag{10.24}$$

are volume fractions of the fluid and solid phases, respectively. Insertion of Equations (10.22) and (10.23) into Equation (10.21) implies that

$$\Psi = \Psi_{\text{solid}} + \mu_0 n + \frac{k_B T}{v}\left(nv \ln \frac{nv}{1+nv} + \chi \frac{nv}{1+nv}\right). \tag{10.25}$$

10.2.3 Derivation of Constitutive Equations

To develop constitutive equations, we apply the method of [14]: the problem of mechanical deformation of a hydrogel subjected to swelling is immersed in a larger class of problems with volume and surface mass uptake (in terminology of [14], pumps injecting solvent are ascribed to each elementary volume of a specimen).

Under quasi-static deformation of a hydrogel, the first Piola–Kirchhoff stress tensor $\mathbf{P}$ satisfies the equilibrium equations

$$\boldsymbol{\nabla}_0 \cdot \mathbf{P} + \mathbf{b} = \mathbf{0} \quad (\text{in } \Omega) \qquad \mathbf{n}_0 \cdot \mathbf{P} = \mathbf{t} \quad (\text{at } \partial\Omega), \tag{10.26}$$

where Ω is an arbitrary domain occupied by the hydrogel in the initial configuration, $\partial\Omega$ is its boundary, $\boldsymbol{\nabla}_0$ is the gradient operator in the initial configuration, $\mathbf{n}_0$ is unit outward normal vector at $\partial\Omega$, $\mathbf{b}$ denotes volume force, and $\mathbf{t}$ is surface traction.

where

$$\Theta = \frac{\partial \Psi}{\partial t} - \frac{2\dot{f}}{3f}\left(J_{e1}\Psi_{,e1} + 2J_{e2}\Psi_{,e2} + 3J_{e3}\Psi_{,e3}\right). \tag{10.36}$$

The molecular incompressibility condition (10.30) establishes a connection between the deformation gradient $\mathbf{F}$ and the rate of injection of solvent R. To account for this dependence, we differentiate Equation (10.30) with respect to time. Keeping in mind that

$$\frac{\mathrm{d}}{\mathrm{d}t}\det \mathbf{F} = (\det \mathbf{F})\mathbf{I} : \mathbf{D},$$

and replacing the derivative of n by means of Equation (10.27), we obtain

$$v(R - \boldsymbol{\nabla}_0 \cdot \mathbf{j}_0) - (\det \mathbf{F})\mathbf{I} : \mathbf{D} = 0. \tag{10.37}$$

Multiplying Equation (10.37) by an arbitrary function Π, integrating over Ω, and performing integration by parts with the help of Equation (10.28), we arrive at

$$\int_\Omega \left[\Pi\left(vR - (\det \mathbf{F})\mathbf{I} : \mathbf{D}\right) + \mathbf{j}_0 \cdot \boldsymbol{\nabla}_0(\Pi v)\right]\mathrm{d}V + \int_{\partial\Omega} \Pi v r \mathrm{d}A = 0. \tag{10.38}$$

Inserting Equations (10.32), (10.35) into Equation (10.31) and adding Equation (10.38), we find that

$$\int_\Omega \Big\{ 2\Big[(\Psi_{,e1}\mathbf{B}_e - J_{e3}\Psi_{,e2}\mathbf{B}_e^{-1}) + (J_{e2}\Psi_{,e2} + J_{e3}\Psi_{,e3})\mathbf{I} -$$

$$(\det \mathbf{F})(\mathbf{T} + \Pi\,\mathbf{I})\Big\} : \mathbf{D}\mathrm{d}V +$$

$$2\int_\Omega \Big[(\Psi_{,p1}\mathbf{B}_p - \Psi_{,p2}\mathbf{B}_p^{-1}) - (\Psi_{,e1}\mathbf{C}_e - J_{e3}\Psi_{,e2}\mathbf{C}_e^{-1})\Big] : \mathbf{d}_p\mathrm{d}V +$$

$$\int_\Omega \Big(\frac{\partial \Psi}{\partial n} + \Pi v - \mu\Big)R\mathrm{d}V + \int_{\partial\Omega} \Big(\frac{\partial \Psi}{\partial n} + \Pi v - \mu\Big)r\mathrm{d}A +$$

$$\int_\Omega \mathbf{j}_0 \cdot \boldsymbol{\nabla}_0\Big(\frac{\partial \Psi}{\partial n} + \Pi v\Big)\mathrm{d}V + \int_\Omega \Theta\mathrm{d}V \leq 0. \tag{10.39}$$

Keeping in mind that $\mathbf{D}, R, r$ are now arbitrary functions (the only connection between them (10.30) is accounted for by means of the function Π), we conclude that the thermodynamic inequality (10.39) is satisfied, provided that (i) the Cauchy stress tensor is given by

$$\mathbf{T} = -\Pi\,\mathbf{I} + \frac{2}{\det \mathbf{F}}\Big[(\Psi_{,e1}\mathbf{B}_e - J_{e3}\Psi_{,e2}\mathbf{B}_e^{-1}) + (J_{e2}\Psi_{,e2} + J_{e3}\Psi_{,e3})\mathbf{I}\Big], \tag{10.40}$$

we arrive at the formula

$$\mathbf{d}_{\mathrm{p}} = P\Big\{\Big[\Psi_{,\mathrm{e}1}\Big(\mathbf{C}_{\mathrm{e}} - \frac{1}{3}J_{1\mathrm{e}}\mathbf{I}\Big) - \Psi_{,\mathrm{e}2}\Big(J_{\mathrm{e}3}\mathbf{C}_{\mathrm{e}}^{-1} - \frac{1}{3}J_{\mathrm{e}2}\mathbf{I}\Big)\Big] - \Big[\Psi_{,\mathrm{p}1}\Big(\mathbf{B}_{\mathrm{p}} - \frac{1}{3}J_{1\mathrm{p}}\mathbf{I}\Big) - \Psi_{,\mathrm{p}2}\Big(\mathbf{B}_{\mathrm{p}}^{-1} - \frac{1}{3}J_{\mathrm{p}2}\mathbf{I}\Big)\Big]\Big\}. \tag{10.46}$$

It follows from Equations (10.8), (10.13), (10.18) that

$$\dot{\mathbf{B}}_{\mathrm{p}} = \mathbf{d}_{\mathrm{p}} \cdot \mathbf{B}_{\mathrm{p}} + \mathbf{B}_{\mathrm{p}} \cdot \mathbf{d}_{\mathrm{p}}.$$

Substitution of Equation (10.46) into this equation yields

$$\dot{\mathbf{B}}_{\mathrm{p}} = 2P\Big[\frac{1}{2}\Psi_{,\mathrm{e}1}(\mathbf{C}_{\mathrm{e}} \cdot \mathbf{B}_{\mathrm{p}} + \mathbf{B}_{\mathrm{p}} \cdot \mathbf{C}_{\mathrm{e}}) - \frac{1}{2}J_{\mathrm{e}3}\Psi_{,\mathrm{e}2}(\mathbf{C}_{\mathrm{e}}^{-1} \cdot \mathbf{B}_{\mathrm{p}} + \mathbf{B}_{\mathrm{p}} \cdot \mathbf{C}_{\mathrm{e}}^{-1})$$
$$-\frac{1}{3}\Big((J_{\mathrm{e}1}\Psi_{,\mathrm{e}1} - J_{\mathrm{e}2}\Psi_{,\mathrm{e}2}) - (J_{\mathrm{p}1}\Psi_{,\mathrm{p}1} - J_{\mathrm{p}2}\Psi_{,\mathrm{p}2})\Big)$$
$$\mathbf{B}_{\mathrm{p}} + \Psi_{,\mathrm{p}2}\mathbf{I} - \Psi_{,\mathrm{p}1}\mathbf{B}_{\mathrm{p}}^{2}\Big]. \tag{10.47}$$

Equations (10.8)–(10.10) and (10.13) imply that

$$\dot{\mathbf{F}}_{\mathrm{e}} = \Big(\mathbf{L} - \frac{\dot{f}}{3f}\mathbf{I}\Big) \cdot \mathbf{F}_{\mathrm{e}} - \mathbf{F}_{\mathrm{e}} \cdot \mathbf{d}_{\mathrm{p}}.$$

Combination of this equation with Equation (10.37) results in

$$\dot{\mathbf{F}}_{\mathrm{e}} = \Big(\mathbf{L} - \frac{\dot{f}}{3f}\mathbf{I}\Big) \cdot \mathbf{F}_{\mathrm{e}} - P\mathbf{F}_{\mathrm{e}} \cdot \Big\{\Big[\Psi_{,\mathrm{e}1}\Big(\mathbf{C}_{\mathrm{e}} - \frac{1}{3}J_{\mathrm{e}1}\mathbf{I}\Big)$$
$$-\Psi_{,\mathrm{e}2}\Big(J_{\mathrm{e}3}\mathbf{C}_{\mathrm{e}}^{-1} - \frac{1}{3}J_{\mathrm{e}2}\mathbf{I}\Big)\Big] - \Big[\Psi_{,\mathrm{p}1}\Big(\mathbf{B}_{\mathrm{p}} - \frac{1}{3}J_{\mathrm{p}1}\mathbf{I}\Big) - \Psi_{,\mathrm{p}2}\Big(\mathbf{B}_{\mathrm{p}}^{-1} - \frac{1}{3}J_{\mathrm{p}2}\mathbf{I}\Big)\Big]\Big\}. \tag{10.48}$$

Given a free energy density (10.20), Equations (10.40), (10.41), (10.47), (10.48) provide stress–strain relations in finite viscoplasticity of hydrogels.

10.3 Simplification of the Constitutive Equations

Our aim now is to perform quantitative investigation of the viscoplastic response of nanocomposite hydrogels in short-term tests whose duration is noticeably lower than the characteristic time for diffusion of solvent. For this purpose, we simplify the constitutive equations in order (i) to make them suitable for fitting experimental data and (ii) to reduce the number of adjustable parameters.

First, we suppose that the strain energy density Ψ_{solid} depends on principal invariants J_{e1}, J_{e3}, and J_{p1} only, and present this function in the form

$$\Psi_{\text{solid}} = W_1(J_{e1}, J_{e3}, t) + W_2(J_{p1}, t), \qquad (10.49)$$

where W_1 denotes mechanical energy stored in individual chains of the equivalent polymer network (this quantity depends on principal invariants of the Cauchy–Green tensor for elastic deformation) and W_2 stands for the energy of interaction between chains and nanoparticles (treated as a function of principal invariants of the Cauchy–Green tensor for plastic deformation). The influence of second principal invariants of the corresponding Cauchy–Green tensors on the mechanical response is disregarded in Equation (10.49).

Substitution of Equations (10.21), (10.24), (10.30), (10.49) into Equation (10.40) implies that

$$\mathbf{T} = -\Pi\,\mathbf{I} + 2\phi_{\text{s}}(w_1\mathbf{B}_{\text{e}} + J_{e3}w_1'\mathbf{I}), \qquad (10.50)$$

where

$$w_1 = \frac{\partial W_1}{\partial J_{e1}}, \qquad w_1' = \frac{\partial W_1}{\partial J_{e3}}. \qquad (10.51)$$

Combination of Equations (10.47–10.49) results in

$$\dot{\mathbf{B}}_{\text{p}} = 2P\left[\frac{1}{2}w_1(\mathbf{C}_{\text{e}} \cdot \mathbf{B}_{\text{p}} + \mathbf{B}_{\text{p}} \cdot \mathbf{C}_{\text{e}}) - \frac{1}{3}(w_1 J_{e1} - w_2 J_{p1})\mathbf{B}_{\text{p}} - w_2\mathbf{B}_{\text{p}}^2\right],$$

$$\dot{\mathbf{F}}_{\text{e}} = \left(\mathbf{L} - \frac{\dot{f}}{3f}\mathbf{I}\right) \cdot \mathbf{F}_{\text{e}} - P\mathbf{F}_{\text{e}} \cdot \left[w_1\left(\mathbf{C}_{\text{e}} - \frac{1}{3}J_{e1}\mathbf{I}\right) - \right.$$
$$\left. w_2\left(\mathbf{B}_{\text{p}} - \frac{1}{3}J_{p1}\mathbf{I}\right)\right] \qquad (10.52)$$

with

$$w_2 = \frac{\partial W_2}{\partial J_{p1}}. \qquad (10.53)$$

Under an arbitrary deformation of a nanocomposite hydrogel subjected to swelling, Equations (10.50)–(10.52) together with Equation (10.42) for chemical potential should be accompanied by the equilibrium equations for the Cauchy stress tensor $\mathbf{T}$ and diffusion Equation (10.27) with $R = 0$,

$$\dot{n} = \boldsymbol{\nabla}_0 \cdot \left(\frac{Dn}{k_B T}\mathbf{F}^{-1} \cdot \boldsymbol{\nabla}_0\mu \cdot \mathbf{F}^{-1}\right).$$

where $\tilde{G}$ stands for an analog of elastic modulus, and $K > 0$, α are material constants. To reduce the number of adjustable parameters, we presume that

$$K = 1, \qquad \alpha = 2$$

under tension, and

$$\alpha = -\frac{1}{2}$$

under retraction.

Insertion of Equations (10.63) and (10.64) into Equations (10.59) and (10.61) results in the stress–strain relation

$$\sigma = GX\phi_s\left[1 - \frac{1}{J}\left(X\left(\frac{k^2}{k_p^2} + 2\frac{k_p}{k}\right) - 3\right)\right]^{-1}\frac{k^3 - k_p^3}{k^2 k_p^2} \tag{10.65}$$

and the kinetic equation for plastic flow

$$\dot{k}_p = S\left\{X\left[1 - \frac{1}{J}\left(X\left(\frac{k^2}{k_p^2} + 2\frac{k_p}{k}\right) - 3\right)\right]^{-1}\frac{k^3 - k_p^3}{k k_p} - \right.$$

$$\left. R\left[1 + K\left(k_p^2 + \frac{2}{k_p} - 3\right)^\alpha\right](k_p^3 - 1)\right\}\frac{|\dot{k}|}{k} \tag{10.66}$$

with

$$S = \frac{PG}{3D}, \qquad R = \frac{\tilde{G}}{G}, \tag{10.67}$$

where $D = |\dot{k}|/k$ stands for strain-rate intensity.

Constitutive Equations (10.65) and (10.66) involve six adjustable parameters: (i) G stands for elastic modulus of an equivalent polymer network, (ii) X characterizes swelling-induced inflation of the network, (iii) J is a measure of extensibility of chains, (iv) S denotes rate of plastic flow, (v) R stands for strength of inter-chain interactions, (vi) K characterizes energy of inter-chain interactions under retraction. These quantities may be affected by composition of a hydrogel, strain rate $\dot{k}$ (due to the neglect of viscoelastic properties associated with rearrangement of polymer network), and deformation program (the energy of interaction W_2 adopts different values under tension and retraction).

Although the number of material constants in the constitutive equations appears to be reasonable compared with conventional models in finite viscoplasticity of polymers, this number can be reduced further for special loading

dried samples to the initial degree of swelling Q, (iii) uniaxial tensile tests on dried–reswollen specimens. Mechanical tests were performed at room temperature with strain rate $\dot{\epsilon} = 0.02 \, \text{s}^{-1}$ ($\epsilon = k - 1$ stands for engineering tensile strain) up to breakage of samples.

Experimental stress–strain diagrams are depicted in Figure 10.1 where engineering stress σ is plotted versus elongation ratio k. To reduce the number of material constants to be found by matching observations, we presume the response of the nanocomposite hydrogels to be merely elastic. By setting $S = 0$ in Equation (10.66), we conclude that $k_{\text{p}} = 1$ and each stress–strain curve is determined by three parameters G, J, X.

We begin with fitting observations on the as-prepared specimen for which Equation (10.68) is fulfilled. To find adjustable parameters G and J, we fix some interval $[0, J^{\circ}]$, where J is located, and divide this interval into $I = 10$ sub-intervals by the points $J^{(i)} = i\Delta J$ with $\Delta J = J^{\circ}/I$ ($i = 0, 1, \ldots, I-1$). For each $J^{(i)}$, Equations (10.65), (10.66) are integrated numerically by the Runge–Kutta method with step $\Delta t = 0.01$. The modulus G is calculated by the least-squares technique from the condition of minimum of the function

$$F = \sum_{n} \left[\sigma^{\text{exp}}(k_n) - \sigma^{\text{num}}(k_n) \right]^2,$$

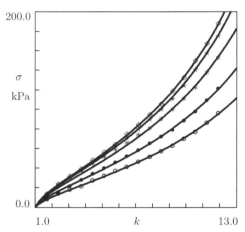

Figure 10.1   Stress σ versus elongation ratio k. Symbols: experimental data in tensile tests on DMAA-NC hydrogel subjected to drying down to various Q_{dry} and subsequent re-swelling up to $Q = 7.2$ ($\circ$ – as-prepared; $\bullet$ – $Q_{\text{dry}} = 3.0$; $*$ – $Q_{\text{dry}} = 1.7$; $\star$ – $Q_{\text{dry}} = 0.8$; $\diamond$ – $Q_{\text{dry}} = 0.06$). Solid lines: results of simulation.

where summation is performed over all elongation ratios k_n at which the observations are reported, $\sigma^{\exp}$ stands for engineering stress measured in the test, and σ^{num} is given by Equation (10.65). The best-fit value of J is found from the condition of minimum of F. Afterwards, the initial interval is replaced with new interval $[J - \Delta J, J + \Delta J]$, and the calculations are repeated.

After finding the best-fit value $G = 48.2$ kPa, we fix this quantity, and match observations on samples subjected to drying-reswelling by means of the above algorithm with adjustable parameters X and J. Given X, we calculate f from Equation (10.60) and plot f and J versus Q_{dry} in Figure 10.2. The data are approximated by the linear equations

$$f = f_0 + f_1 Q_{\text{dry}}, \qquad J = J_0 + J_1 Q_{\text{dry}} \qquad (10.69)$$

with coefficients calculated by the least-squares technique. Following [39], the strong (by twice) reduction in f induced by drying and re-swelling is attributed to rearrangement of the secondary network (a house-of-cards structure formed by clay platelets).

10.4.2 As-Prepared Poly(Dimethylacrylamide)–Silica Hydrogels

We proceed with the analysis of observations on polydimethylacrylamide–silica (DMAA-Si) hydrogels manufactured by free-radical polymerization of N,N-dimethylacrylamide in aqueous suspensions of silica nanoparticles

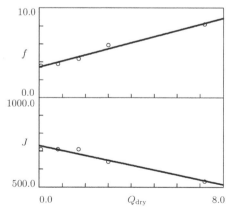

Figure 10.2 Parameters f and J versus solvent content after drying Q_{dry}. Circles: treatment of observations on DMAA-NC hydrogels. Solid lines: approximation of the data by Equation (10.69).

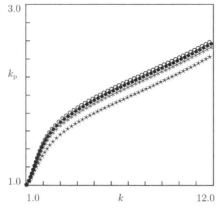

Figure 10.5 Elongation ratio for plastic deformation k_p versus elongation ratio k. Symbols: results of simulation for tensile tests on DMAA-Si hydrogels with $\phi_p = 142$ g/L and various ϕ_f g/L ($\circ$ – 710.5; $\bullet$ – 284.4; $*$ – 142.7; $\star$ – 71.4).

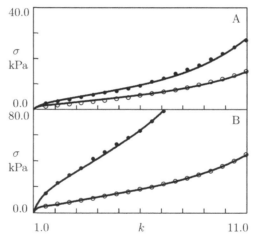

Figure 10.6 Stress σ versus elongation ratio k. Symbols: experimental data in tensile tests with various strain rates $\dot{\epsilon}$ s^{-1} ($\circ$ – 0.06; $\bullet$ – 0.6) on DMAA-Si hydrogels with $\phi_p = 142.2$ g/L and $\phi_f = 142.7$ g/L (A), $\phi_f = 710.5$ g/L (B). Solid lines: results of simulation.

Evolution of elastic moduli with strain rate $\dot{\epsilon}$ is illustrated in Figure 10.7 where G and $\tilde{G}$ are depicted versus duration of loading $t' = (k_{max} - 1)/\dot{\epsilon}$ with $k_{max} = 11$. For comparison, observations in tensile relaxation tests with strain $\epsilon = 0.5$ are also presented (in semi-logarithmic coordinates with $\log = \log_{10}$). Following [40], relaxation curves are plotted in the form $\tilde{\sigma}(t')$

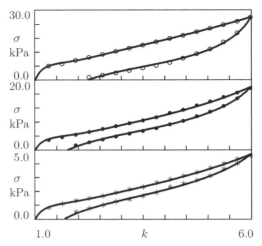

Figure 10.8 Stress σ versus elongation ratio k. Symbols: experimental data in cyclic tests on DMAA-Si hydrogels with $\phi_p = 142$ g/L and various ϕ_f g/L ($\circ$ – 710.5; $\bullet$ – 284.4; $*$ – 142.7). Solid lines: results of simulation.

is found by matching observations on hydrogel with $\phi_f = 710.5$ g/L and used without changes to approximate observations on other specimens. Numerical analysis shows that the best-fit values of J are slightly lower and the best-fit values of G are slightly higher than those found by matching observations in Figure 10.3.

The effect of concentration of solid phase ϕ_s on coefficients $\tilde{G}$ and S found by matching observations under retraction is illustrated in Figure 10.9. The data are approximated by the equations

$$\log \tilde{G} = \tilde{G}_0 - \tilde{G}_1\phi_s, \qquad \log S = S_0 + S_1\phi_s, \qquad (10.71)$$

with coefficients calculated by the least-squares method. Figure 10.9 shows that the energy of inter-chain interactions $\tilde{G}$ decreases and rate of plastic deformation S increases with concentration of solid phase.

To examine the kinetics of plastic flow under cyclic loading, integration of the stress–strain relations is conducted with the adjustable parameters found by matching observations in Figure 10.8. Results of simulation are presented in Figure 10.10. The following conclusions are drawn: (i) under tension, plastic elongation ratio k_p increases monotonically with k and remains practically independent of clay content (in agreement with the data reported in Figure 10.5), (ii) under retraction, k_p grows pronouncedly at the initial

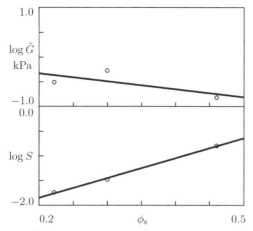

Figure 10.9 Parameters $\tilde{G}$ and S versus concentration of solid phase ϕ_s. Circles: treatment of observations under retraction in cyclic tests on DMAA-Si hydrogels with $\phi_p = 142$ g/L and various ϕ_f g/L. Solid lines: approximation of the data by Equation (10.71).

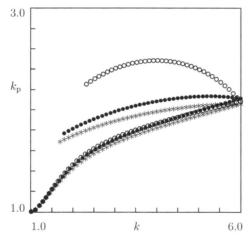

Figure 10.10 Elongation ratio for plastic deformation k_p versus elongation ratio k. Symbols: results of simulation for cyclic tests on DMAA-Si hydrogels with $\phi_p = 142$ g/L and various ϕ_f g/L ($\circ$ – 710.5; $\bullet$ – 284.4; $*$ – 142.7).

stage of unloading (when k remains in the vicinity of k_{max}), reaches its maximum (plastic overshoot), and decreases afterwards, (iii) intensity of plastic overshoot ($k_{p\,max}$) and residual strain (k_p at the instant when σ vanishes) increase strongly (by twice) with nanoclay content.

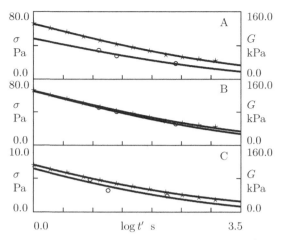

Figure 10.14 Dimensionless stress $\tilde{\sigma}$ ($\star$) versus relaxation time t' and modulus G versus loading time t'. Symbols: observations on PAM-NC hydrogels with $\phi_p = 200$ g/L, $\phi_f = 20$ g/L (A), $\phi_p = 250$ g/L, $\phi_f = 20$ g/L (B), $\phi_p = 100$ g/L, $\phi_f = 40$ g/L (C). Stars: experimental data in tensile relaxation test with strain $\epsilon = 0.005$. Circles: treatment of experimental data in tensile tests with various strain rates. Solid lines: results of simulation.

them with those on samples with $\phi_p = 250$ g/L, $\phi_f = 20$ g/L). According to Figure 10.14, an increase in elastic modulus with strain rate revealed by fitting observations in tensile tests may be attributed to the viscoelastic response of hydrogels (in agreement with the conclusion drawn from Figure 10.7).

To assess the influence of strain rate on the viscoplastic flow, parameter S is plotted versus $\dot{\epsilon}$ in Figure 10.15. The data are approximated by the equation

$$\log S = S_0 + S_1 \log \dot{\epsilon}, \tag{10.72}$$

where the coefficients are calculated by the least-squares technique. Equation (10.72) demonstrates that S increases monotonically with $\dot{\epsilon}$, but the rate of growth is strongly sub-linear: $S \sim \dot{\epsilon}^\beta$ with $\beta \approx 0.2$.

Given G and R, we calculate $\tilde{G}$ from Equation (10.67) and plot this quantity versus strain rate $\dot{\epsilon}$ in Figure 10.16. The data are approximated by the equation

$$\log \tilde{G} = \tilde{G}_0 + \tilde{G}_1 \log \dot{\epsilon} \tag{10.73}$$

with coefficients calculated by the least-squares technique. Figure 10.16 demonstrates a pronounced difference between PAM-NC hydrogels (for which $\tilde{G}$ increases strongly with strain rate) and DMAA-Si hydrogels (for which this parameter is independent of $\dot{\epsilon}$).

To examine the effect of concentration of solid phase ϕ_s on the mechanical response of nanocomposite hydrogels under cyclic loading, we approximate stress–strain curves under tension with constant strain rate $\dot{\epsilon} = 0.083$ s^{-1} up to maximum elongation ratio $k_{max} = 17$ followed by retraction with the same strain rate down to the zero minimum stress $\sigma_{min} = 0$. Observations on specimens with $\phi_f = 20$ g/L, $\phi_p = 200$ and 250 g/L are reported in Figure 10.17, and those on samples with $\phi_p = 100$ g/L, $\phi_f = 30$ and 40 g/L are presented in Figure 10.18 together with results of simulation. Adjustable parameters in the stress–strain relations are found by fitting each set of data separately by means of the same algorithm that was employed to match observations in Figure 10.8. As the stress–strain diagrams under tension in Figures 10.17, 10.18 differ from those reported in Figures 10.11–10.13, they are approximated by using four parameters G, J, S, R. Afterwards, unloading paths of the stress–strain curves are fitted with the help of three coefficients K, S, R.

Adjustable parameters determined by matching observations under tension and retraction are reported in Figures 10.19, 10.20, where they are plotted versus concentration of solid phase ϕ_s. The data are approximated by the equations

$$\log G = G_0 + G_1 \phi_s, \qquad \log \tilde{G} = \tilde{G}_0 + \tilde{G}_1 \phi_s, \qquad \log S = S_0 + S_1 \phi_s,$$

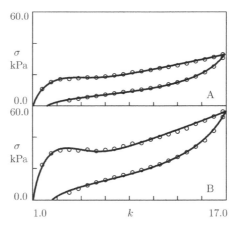

Figure 10.17 Stress σ versus elongation ratio k. Circles: experimental data in cyclic tensile tests on PAM-NC hydrogels with $\phi_f = 20$ g/L and $\phi_p = 200$ g/L (A), $\phi_p = 250$ g/L (B). Solid lines: results of simulation.

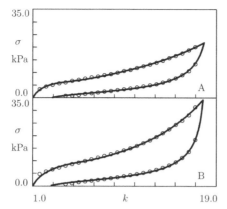

Figure 10.18 Stress σ versus elongation ratio k. Circles: experimental data in cyclic tests on PAM-NC hydrogels with $\phi_n = 100$ g/L and $\phi_f = 30$ g/L (A), $\phi_f = 40$ g/L (B). Solid lines: results of simulation.

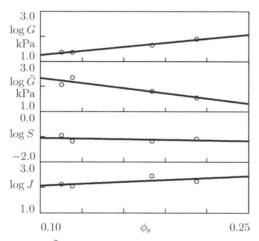

Figure 10.19 Parameters $G, \tilde{G}, S, J$ versus concentration of solid phase ϕ_s. Circles: treatment of observations under tension in cyclic tests on PAM-NC hydrogels. Solid lines: approximation of the data by Equation (10.74).

$$\log J = J_0 + J_1 \phi_s \qquad (10.74)$$

under tension and

$$\log \tilde{G} = \tilde{G}_0 + \tilde{G}_1 \phi_s, \qquad \log S = S_0 + S_1 \phi_s, \qquad \log K = K_0 + K_1 \phi_s \qquad (10.75)$$

time-dependent phenomena in nanocomposite hydrogels, rearrangement of the networks of flexible chains and nanoparticles should be modeled explicitly (the latter leads, however, to a noticeable increase in the number of material constants).

Comparison of experimental data under cyclic deformation with results of numerical simulation demonstrates that elongation ratio for plastic deformation k_p grows monotonically under tension and reveals a non-monotonic behavior under retraction: k_p increases, reaches its maximum value (plastic overshoot), and decreases afterwards (Figures 10.10 and 10.21). Intensity of the overshoot increases with concentration of solid phase, but the rate of its growth depends strongly on type of nanofiller.

10.5 Concluding Remarks

A constitutive model is developed in finite viscoplasticity of nanocomposite hydrogels under an arbitrary deformation with finite strains. A hydrogel is treated as a two-phase medium composed of a solid phase (polymer network reinforced with nanoparticles) and a fluid phase (solvent). Transport of solvent through a hydrogel is treated as its diffusion governed by the gradient of chemical potential.

Constitutive equations are derived by means of the free-energy imbalance equation. The free energy of a nanocomposite hydrogel equals the sum of strain energy density of the solid phase and the energy of mixing of the solid phase with solvent.

The solid phase is modeled as an isotropic compressible viscoplastic medium, whose deformation gradient is split into the product of deformation gradients for elastic deformation, plastic deformation, and deformation induced by swelling (characterized by the coefficient of inflation of the polymer network). Strain energy density of the solid phase equals the sum of the stored mechanical energy and the energy of interaction between chains and nanoparticles.

The constitutive equations involve (i) stress–strain relation, (ii) flow rule for plastic deformation, and (iii) diffusion equation for solvent. These relations are accompanied by equations for mechanical equilibrium and appropriate boundary conditions.

The model is applied to the analysis of rapid deformation (the rate of loading exceeds strongly the rate of solvent diffusion) of (i) nanocomposite hydrogels subjected to drying and subsequent re-swelling, and (ii) as-prepared nanocomposite hydrogels.

[7] Buenger, D., Topuz, F., Groll, J., 'Hydrogels in sensing applications', Progr. Polym. Sci., 37, 1678–1719, 2012.

[8] Qian, Z.-Y., Fu, S.-Z., Feng, S.-S., 'Nanohydrogels as a prospective member of the nanomedicine family', Nanomedicine, 8, 161–164, 2013.

[9] Higuchi, A., Ling, Q.-D., Chang, Y., Hsu, S.-T., Umezawa, A., 'Physical cues of biomaterials guide stem cell differentiation fate', Chem. Rev., 113, 3297–3328, 2013.

[10] Discher, D.E., Mooney, D.J., Zandstra, P.W., 'Growth factors, matrices, and forces combine and control stem cells', Science, 324, 1673–1677, 2009.

[11] Song, M.J., Dean, D., Knothe Tate, M.L., 'Mechanical modulation of nascent stem cell lineage commitment in tissue engineering scaffolds', Biomaterials, 34, 5766–5775, 2013.

[12] Pelaez, D., Huang, C.-Y.C, Cheung, H.S., 'Cyclic compression maintains viability and induces chondrogenesis of human mesenchymal stem cells in fibrin gel scaffolds', Stem Cells Develop., 18, 93–102, 2009.

[13] Meyer, E.G., Buckley, C.T., Steward, A.J., Kelly, D.J., 'The effect of cyclic hydrostatic pressure on the functional development of cartilaginous tissues engineered using bone marrow derived mesenchymal stem cells', J. Mech. Behav. Biomed. Mater., 4, 1257–1265, 2011.

[14] Hong, W., Zhao, X., Zhou, J., Suo, Z., 'A theory of coupled diffusion and large deformation in polymeric gels', J. Mech. Phys. Solids, 56, 1779–1793, 2008.

[15] Hong, W., Liu, Z., Suo, Z., 'Inhomogeneous swelling of a gel in equilibrium with a solvent and mechanical load', Int. J. Solids Struct., 46, 3282–3289, 2009.

[16] Duda, F.P., Souza, A.C., Fried, E., 'A theory for species migration in finitely strained solid with application to polymer network swelling', J. Mech. Phys. Solids, 58, 515–529, 2010.

[17] Chester, S.A., Anand, L., 'A coupled theory of fluid permeation and large deformations for elastomeric materials', J. Mech. Phys. Solids, 58, 1879–1906, 2010.

[18] Baek, S., Pence, T.J., 'Inhomogeneous deformation of elastomer gels in equilibrium under saturated and unsaturated conditions', J. Mech. Phys. Solids, 59, 561–582, 2011.

[19] Yan, H., Jin, B., 'Influence of microstructural parameters on mechanical behavior of polymer gels', Int. J. Solids Struct., 49, 436–444, 2012.

[20] Drozdov, A.D., Christiansen, J.deC., 'Constitutive equations in finite elasticity of swollen elastomers', Int. J. Solids Struct., 50, 1494–1504, 2013.

[21] Drozdov, A.D., Christiansen, J.deC., 'Stress–strain relations for hydrogels under multiaxial deformation', Int. J. Solids Struct., 50, 3570–3585, 2013.

[22] Lucantonio, A., Nardinocchi, P., Teresi, L., 'Transient analysis of swelling-induced large deformations in polymer gels', J. Mech. Phys. Solids, 61, 205–218, 2013.

[23] Zhao, X., Huebsch, N., Mooney, D.J., Suo, Z., 'Stress–relaxation behavior in gels with ionic and covalent crosslinks', J. Appl. Phys., 107, 063509, 2010.

[24] Chester, S.A., 'A constitutive model for coupled fluid permeation and large viscoelastic deformation in polymeric gels', Soft Matter, 8, 8223–8233, 2012.

[25] Wang, X., Hong, W., 'A visco-poroelastic theory of polymeric gels', Proc. Roy. Soc. A, 468, 3824–3841, 2012.

[26] Strange, D.G.T., Fletcher, T.L., Tonsomboon, K., Brawn, H., Zhao, X., Oyen, M.L., 'Separating poroviscoelastic deformation mechanisms in hydrogels', Appl. Phys. Lett., 102, 031913, 2013.

[27] Baumberger, T., Caroli, C., Martina, D., 'Solvent control of crack dynamics in a reversible hydrogel', Nature Mater., 5, 552–555, 2006.

[28] Seitz, M.E., Martina, D., Baumberger, T., Krishnan, V.R., Hui, C.-Y., Shull, K.R., 'Fracture and large strain behavior of self-assembled triblock copolymer gels', Soft Matter, 5, 447–456, 2009.

[29] Kundu, S., Crosby, A.J., 'Cavitation and fracture behavior of polyacrylamide hydrogels', Soft Matter, 5, 3963–3968, 2009.

[30] Wang, X., Hong, W., 'Delayed fracture in gels', Soft Matter, 8, 8171–8178, 2012.

[31] Zhang, J., An, Y., Yazzie, K., Chawla, N., Jiang, H., 'Finite element simulation of swelling-induced crack healing in gels', Soft Matter, 8, 8107–8112, 2012.

[32] Korchagin, V., Dowbow, J., Stepp, D., 'A theory of amorphous viscoelastic solids undergoing finite deformations with application to hydrogels', Int. J. Solids Struct., 44, 3973–3997, 2007.

[33] Wang, X., Hong, W., 'Pseudo-elasticity of a double network gel', Soft Matter, 7, 8576–8581, 2011.

[34] Zhao, X., 'A theory for large deformation and damage of interpenetrating polymer networks', J. Mech. Phys. Solids, 60, 319–332, 2012.

ATMP comprise a variety of novel therapeutic strategies, including Cell Therapy, Gene Therapy and Tissue Engineered Medicinal Products (CTMP, GTMP, TEP, respectively).

In Europe, as in most countries in the world, ATMP are considered as medicines, therefore they are covered by the pharmaceutical legislation.

11.2 European Regulatory Frame for ATMP

ATMP are covered by the European legal frame for medicinal products through the EU Regulation 1394/2007 [1]. The current EU definition of GTMP and of CTMP is contained in the new Annex 1 that has been recently issued amending Directive 2001/83/EC [2], while the definition of TEP as well as of combined ATMP is contained in the EU Regulation 1394/2007.

GTMP are defined as those biological medicinal products that contain or consist of a recombinant nucleic acid used in or administered to human beings with a view to regulating, repairing, replacing, adding or deleting a genetic sequence; in addition, their therapeutic, prophylactic or diagnostic effects relate directly to the recombinant nucleic acid sequence they contain, or to the product of genetic expression of that sequence. Therefore, medicinal products such as viral or non viral vectors, plasmid or bacterial vectors, recombinant oncolytic viruses, genetically modified cells, cancer immune-therapeutics (so called "cancer vaccines") are considered GTMP if they fulfil the current definition. It should be noted that, according to new Annex 1, vaccines for infectious diseases are no longer considered GTMP [2].

Somatic CTMP are defined as those biological medicinal products that: i) contain or consist of cells or tissues that have been subject to substantial manipulation so that their biological characteristics, physiological functions or structural properties relevant for the intended clinical use have been altered, or ii) contain or consist of cells or tissues that are not intended to be used for the same essential function(s) in the recipient and the donor, and iii) are presented as having properties for, or are used in or administered to human beings with a view to treating, preventing or diagnosing a disease through the pharmacological, immunological or metabolic action of its cells or tissues [2].

TEP are defined as those biological medicinal products that contain or consist of engineered cells or tissues; and are presented as having properties for, or are used in or administered to humans with a view of regenerating, repairing or replacing a human tissue. Cells/tissues shall be considered 'engineered' if they have been subject to substantial manipulation or are not

Development of medicines includes basically two steps: preclinical and clinical development. The first part includes designing the substance/molecule that represents the medicine, proving its activity and testing its safety in a preclinical model, while the latter part includes translating those results into human subjects by means of clinical trials.

As for any other medicine, in Europe clinical trials with ATMP are covered by the Directive 2001/20/EC [6], that stipulates that for all medicines, clinical trial approval is the responsibility of Competent Authority in each European Member State (MS). Therefore clinical development takes place at national level. For any given clinical trial to be performed in a given MS, the EU Member State performs a separate evaluation and authorization procedure. This is also the case when a multinational trial is to be carried out, which may represent a difficult task if reviewers have divergent opinion in the different EU MS on the same clinical trial proposal. Procedures and initiatives have been put in place by EMA and national Competent Authorities to decrease the chance that such divergences occur and to facilitate an efficient translation of research discoveries into effective ATMP.

EMA and national Competent Authorities are also concerned that patients are offered SC-based medicinal products only under controlled conditions, such as e.g. in clinical trials or if market authorized. To ensure patients' safety, SC-based medicinal product development should comply with the highest standards, as for any investigational medicinal product, under the supervision of statutory regulatory bodies.

Thus a number of guidance documents (guidelines or reflection papers) describing quality, preclinical and clinical requirements for CPMP/TEP as well as for GTMP, have been produced by EMA to help applicants in developing their products. All guidance documents are available through the EMA website [11].

The main guidance documents for CTMP/TEP is the guideline on human cell-based medicinal products [12], while for GTMP it is the guideline on quality, preclinical and clinical aspects of gene transfer medicinal products [13].

Other guidance documents are available, covering aspects such as for example: risk based approach [14], chondrocyte-based CTMP [15], potency testing of cell-based immunotherapy medicinal products for the treatment of cancer [16], clinical aspects related to TEP [17], genetically modified cells [18], risk of germ-line transmission for GTMP [19], long term follow up for GTMP [20], environmental risk assessment for GTMP [21], non-clinical studies required before first clinical use of GTMP [22].

CAT has recently produced a reflection paper [23] that covers specifically stem cell (SC)-based ATMP. This reflection paper addresses all medicinal products that are presented for marketing authorization and that use any types of SC as starting material, regardless of SC differentiation status in the final product.

11.3 Stem Cell-Based ATMP

SC therapy holds the promise to treat degenerative diseases, cancer and repair of damaged tissues for which there are currently no or limited therapeutic options. SC-based ATMP in general can be classified as CTMP or TEP; if they are genetically modified, they are classified as GTMP. When they contain a MD, they are considered as combined ATMP.

SC-based ATMP can be obtained from adult stem cells or pluripotent stem cells, such as mesenchymal/stromal stem cells (MSC), hematopoietic stem cells (HSC), tissue-specific progenitor cells.

They can be also prepared from human embryonic stem cells (hESC) or induced pluripotent stem cells (iPSC).

Although stem cells share the same principal characteristics of potential for self-renewal and differentiation, SC-based medicinal products do not constitute a homogeneous class. Instead, they represent a spectrum of different cell-based products for which there is a variable degree of scientific knowledge and clinical experience available. For example, while MSCs or HSCs have been more extensively used for therapeutic purposes, this is not the case for hESCs or iPSCs.

Despite their clinical potential, SC-based ATMP bear also potential risks, that require a thorough evaluation before clinical use. Different levels of risk can be associated with specific types of SC. For example, the risk profile associated with iPSCs is expected to be different from those of adult SC (e.g. MSCs or HSCs) for which a substantial amount of clinical experience has already been gained.

The risk profile of SC-based ATMP depends on many risk factors, such as for example the type of stem cells, their differentiation status and proliferation capacity, *in vitro* manipulation steps, the route of administration and the intended site for clinical effect, the irreversibility of treatment or on the other hand the risk of cell loss, the long-term survival of engrafted cells.

The risks so far identified in clinical experience or the potential risks (i.e. those observed in animal studies) include tumour formation, unwanted immune responses and the transmission of adventitious agents.

Even though animal models reflecting the addressed disease would be ideal, in practice this can be prevented by several limitations. In fact, not only the relevant model strain may not be available, but also large animal models may be preferable for studying surgically implanted products, or for long-term evaluation of tissue regeneration and repair or in those situations where the animal size, organ physiology or immune system is relevant for the clinical effect (e.g. regeneration of tissue). Very likely, more than one animal species or strains might be needed. *In vitro* models may also provide additional and/or alternative ways to address some specific aspects.

Ideally, the human cell product should be used, thus requiring immune-suppressed animals. However, for studying aspects such as e.g. persistence or functionality, homologous animal models might provide the most relevant system, even though predictiveness of such a model may be limited because of the still limited knowledge of the similarity between animal and human SC differentiation processes.

11.6 Clinical Issues for Stem Cell-Based Product Development

Ideally, nonclinical evidence on the proof-of-concept and safety of the SC-based product is expected before administration to humans.

In practice, as discussed above, there may be cases where sufficient nonclinical data cannot be obtained. In such cases the evidence should be generated in clinical studies by including additional end points for efficacy and safety, for instance to address the effect of an altered microenvironment (e.g. by inflammation, ischemia).

Clinical studies should evaluate different aspects including proof-of-concept, mode of action, dose finding, biodistribution, persistence, ectopic presence of SC product, safety and long term efficacy.

The mode of action of a SC- based product may be directly dependent on the stem cell population, on molecules secreted by the cells or on their engraftment in the host tissue.

Studies to follow the cells during the clinical studies, i.e. to assess biodistribution of SC product, may be important, depending on the SC product risk profile and its mode/site of administration. Due to a number of circulating SC higher than in physiological condition, abnormal distribution may occur leading to ectopic engraftment in non-target tissues.

On the other hand, such studies may be very imposing on the patients and in practice are hampered by the lack of suitable non invasive tracers.

[7] EU Regulation 1394/2007, art 17

[8] EU Regulation 1394/2007, art 18

[9] EU Regulation 1394/2007, art 14

[10] EU Regulation 1394/2007, art 15

[11] http://www.ema.europa.eu

[12] CHMP/410869/2006 Guideline on human cell-based medicinal products

[13] CPMP/BWP/3088/99 Note for Guidance on quality, preclinical and clinical aspects of gene transfer medicinal products

[14] CAT/CPWP/686637/2011 Guideline on Risk-based approach according to Annex I, part IV of Directive 2001/83/EC applied to Advanced Therapy Medicinal Products

[15] EMA/CAT/CPWP/568181/2009 Reflection paper on *in-vitro* cultured chondrocyte containing products for cartilage repair of the knee

[16] EMEA/CHMP/BWP/271475/2006 Guideline on potency testing of cell based immunotherapy medicinal products for the treatment of cancer

[17] EMA/CAT/CPWP/573420/2009 Draft reflection paper on clinical aspects related to tissue- engineered products

[18] EMA/CAT/GTWP/671639/2008 Guideline on quality, non-clinical and clinical aspects of medicinal products containing genetically modified cells

[19] EMEA/273974/2005 Guideline on non-clinical testing for inadvertent germline transmission of gene transfer vectors

[20] EMEA/CHMP/GTWP/60436/2007 Guideline on follow up of patients administered with gene therapy medicinal products

[21] EMEA/CHMP/GTWP/125491/2006 Guideline on scientific requirements for the environmental risk assessment of gene therapy medicinal products

[22] EMEA/CHMP/GTWP/125459/2006 Guideline on the non-clinical studies required before first clinical use of gene therapy medicinal products

[23] EMA/CAT/571134/2009 Reflection paper on stem cell-based medicinal products

[24] Directive 2004/23/EC, Official Journal of the European Communities, L102/48, 7/4/2004

[25] Directive 2006/17/EC, Official Journal of the European Communities, L38/40, 9/2/2006

[26] European Pharmacopoeia chapter 5.1.7 Viral Safety

Index

Lightning Source UK Ltd.
Milton Keynes UK
UKOW07n2100110115

244311UK00002B/13/P